This series brings together work that takes cognitive science in new directions. For many years, philosophical contributions to the field of cognitive science came primarily from theorists with commitments to physical reductionism, neurocentrism, and a representationalist model of the mind. However, over the last two decades, a rich literature that challenges these traditional views has emerged. According to so-called '4E' approaches, the mind is embodied, embedded, enactive, and extended. Cognition, emotion, and consciousness are not best understood as comprised of brain-bound representational mechanisms, but rather as dynamic, embodied, action-oriented processes that sometimes extend beyond the human body. Such work often draws from phenomenology and dynamic systems theory to rethink the nature of cognition, characterizing it in terms of the embodied activity of an affectively attuned organism embedded in its social world. In recent years, theorists have begun to utilize 4E approaches to investigate questions in philosophy of psychiatry, moral psychology, ethics, and political philosophy. To foster this growing interest in rethinking traditional philosophical notions of cognition using phenomenology, dynamic systems theory, and 4E approaches, we dedicate this series to "New Directions in Philosophy and Cognitive Science."

If you are interested in the series or wish to submit a proposal, please contact Amy Invernizzi, amy.invernizzi@palgrave-usa.com.

New Directions in Philosophy and Cognitive Science

Series Editor
Michelle Maiese
Emmanuel College
Boston, MA, USA

Grant Gillett • Walter Glannon

The Neurodynamic Soul

palgrave
macmillan

Grant Gillett
University of Otago
Dunedin, New Zealand

Walter Glannon
University of Calgary
Calgary, AB, Canada

ISSN 2946-2959   ISSN 2946-2967   (electronic)
New Directions in Philosophy and Cognitive Science
ISBN 978-3-031-44950-5   ISBN 978-3-031-44951-2   (eBook)
https://doi.org/10.1007/978-3-031-44951-2

Cover credit - Ali Kahfi/Getty Images

This Palgrave Macmillan imprint is published by the registered company Springer Nature Switzerland AG.
The registered company address is: Gewerbestrasse 11, 6330 Cham, Switzerland

Paper in this product is recyclable.

Preface and Acknowledgments

In *The Neurodynamic Soul*, we develop a conception of the soul as an integrated set of neural, mental, and bodily functions and capacities that enable flexible and adaptive thought and behaviour. The book combines contemporary neuroscience and philosophy of mind to examine the different respects in which we plan, act, and interact with others. It is distinctive in providing a more robust account of the soul, or mind, than other philosophical accounts by framing it in neuroscientific terms that do not explain it away but explain it as something that interacts with the brain in responding to the natural and social environment. It is also distinctive in relying on both Continental and Anglo-American Analytic philosophy to show how human subjects are embodied and embedded in physical, social, and cultural contexts, and how they use discursive and other abilities to facilitate their engagement with them. The book includes indigenous perspectives on language and communication in the USA, Canada, Australasia, and the Pacific to provide a socially and culturally richer conception of being-in-the-world.

The book is intended for a multidisciplinary audience. This includes philosophers from both Continental and Analytic traditions, cognitive and clinical neuroscientists, psychologists, cultural anthropologists, and informed lay readers.

We thank our editor at Palgrave Macmillan, Amy Invernizzi, for supporting and directing this project. We are also grateful to the two reviewers of the book proposal and the two reviewers of the entire manuscript for their thoughtful and constructive comments. Grant Gillett: special thanks to Shirley, Matthew, Lizzie and Rachel Gillett, as well to the Bioethics

Centre and Philosophy Department at the University of Otago. Walter Glannon: special thanks to Teresa Yee-Wah Yu.

A section of Chap. 2 includes material from "The neurodynamics of free will", *Mind and Matter* 18 (2020): 159–173. We gratefully acknowledge permission from Imprint Academic to publish it in this book.

Grant Gillett is Professor Emeritus of Biomedical Ethics and Philosophy at the University of Otago, New Zealand. He received an MBChB in medicine from the University of Auckland and an MSc in psychology and DPhil in Philosophy from Oxford University. He was a practicing neurosurgeon at Dunedin Hospital from 1988 to 2010.

Walter Glannon is Professor Emeritus of Philosophy at the University of Calgary. He received a PhD in philosophy from Yale University.

Dunedin, New Zealand   Grant Gillett
Calgary, AB, Canada   Walter Glannon

CONTENTS

CHAPTER 1

Introduction

This book is an analysis and discussion of the soul as a psychophysical process and its role in mental representation, meaning, understanding, and agency. Based in part on Aristotle's model of the psyche, it is informed by the current neuroscientific model of the brain as a dynamic organ in which patterns of neural oscillation and synchronization are shaped by biological, social, and cultural factors. These patterns, and their response to these factors, promote an open-ended homeostasis within the human organism and its ability to respond and adapt to, engage with, and flourish in the world.

According to this model of the soul, there can be no mind without brain and no brain without mind (Zeman, 2020). The mental and neural are interacting systems of a human organism necessary for flexibility and adaptability to changing circumstances. These systems constrain and are constrained by each other in a multi-level unified system grounded in space and time. This is in contrast to cognitive science models of the mind or brain as a mechanism of informational input and predictable behavioural output and in which mental states and actions can be explained in physically determined terms. Instead, the brain is an evolving organ whose functions cannot be completely predicted because of its ability to change in response to the environment. The brain's ability to change provides the subject with the sensorimotor, cognitive, affective, and volitional capacities necessary to respond appropriately to the world. While a philosophical

G. Gillett, W. Glannon, *The Neurodynamic Soul*, New Directions in
Philosophy and Cognitive Science,
https://doi.org/10.1007/978-3-031-44951-2_1

account of thought and behaviour requires grounding in neurobiology, they cannot be reduced to neuronal ensembles and how they are excited or inhibited because thought and behaviour involve more than neural features alone.

The book combines contemporary theory on dynamic neuroscience with philosophy of mind and moral philosophy in addressing fundamental issues about human existence. It elucidates how Aristotelian concepts of *psuche* (soul, mind), *phronesis* (practical wisdom), *arete* (excellence), and *eudaimonia* (flourishing) form a holistic model of human behaviour. These concepts can be integrated into an explanation of how humans express these properties in planning, acting, and interacting with others.

It provides a more robust account of the soul, or mind, than other philosophical accounts by framing it in neuroscientific terms that do not explain it away but explain it as something that is shaped by how it interacts with the brain and responds to the natural and social milieu. The book uses a neurodynamic model that relies on both Continental and Anglo-American Analytic philosophy. It describes and explains human subjects as embodied and embedded in the physical, social, and cultural world rather than separated from it, and who use discursive and other abilities to facilitate their engagement with it. It includes indigenous perspectives to give a richer texture to the context of action. In addition, the book is different from historical and contemporary work in normative ethics by defining 'good' as a property of individual and collective adaptability to the world, and how this can promote human excellence and flourishing.

It is instructive to place our conception of the neurodynamic soul within a historical framework. While many philosophers have proposed and defended different conceptions of the soul, we focus on the key *dramatis personae* to provide the context. In the *Phaedo*, Plato argues that the soul, or psyche, is an immaterial substance that survives the demise of the material body (Plato, 1961). The immortal soul consists of a conscious rational faculty that essentially defines persons. Plato's mature tripartite psychology in the later *Republic* divides the soul into rational, spirited (emotional), and appetitive (desiderative) parts. The soul of the just person is one in which the rational part controls the spirited and appetitive parts. Although he conceives of the soul as an immaterial substance, Plato identifies it as what animates the body and guides human behaviour (Broadie, 2001; Karasmanis, 2006).

In his discussion of substance in Book I of the *Physics* and Book VII of the *Metaphysics*, Aristotle argues that form is what unifies some matter into a single object (Aristotle, 1984). Matter becomes the substance or object it is by the form it is. In the *De Anima*, Aristotle defends a hylomorphic conception of a human as a unity of matter (*hyle*) and immaterial form (*morphe*). The soul is the form of the material body in the sense that it is the main organizing or informing principle of the body. The soul is the first actuality of the body of a human being, or what makes it the kind of thing it is. Aristotle describes three degrees or levels of soul (*anima*) in all living things. The first level involves growth and nutrition, the second involves locomotion and perception, and the third involves intellect or thought (*De Anima*, 413a23). These correspond to the nutritive soul (plants), the sensitive soul (all animals), and the rational soul (humans). These levels are nested in the sense that any living thing that has a higher level of soul also contains lower levels. All three levels are present only in humans. In some passages of the *De Anima*, however, Aristotle suggests that the rational soul, or intellect (*oi nou*), is the subject of mental states and not identical to the hylomorphic composite (429a10–11). It is not clear whether the rational soul is an independent or emergent but nested feature of the composite. Aristotle's account suggests that it is more than a unity of matter and form. Drawing upon the pre-Socratic philosopher Democritus, the Epicureans defended a materialistic concept of the soul as a collection of atoms that dissolve at death and are not reconstituted thereafter in successor beings (Warren, 2009). But atomism had little influence on post-Aristotelian philosophy of mind.

Influenced by Aristotle, Aquinas also distinguishes three types of soul corresponding to increasingly complex levels of organization in living things (Aquinas, 1973–1980, pt. 1, question 75, article 5). The nutritive, or vegetative, soul has the lowest level of biological organization and is found in the human zygote and embryo. It is similar to the level of biological organization in plants. The sensitive, or animal, soul displays a higher level of organization and is manifest in the capacity for sensory experience. It is present in all sentient animals. The highest level of organization is found in the rational soul (*anima intellectiva*) and consists in the capacity generally for mentality and specifically for rational thought. The rational soul distinguishes humans from other animals.

According to Descartes' substance dualism, there are two kinds of substance: matter, whose essential property is that it is spatially extended (*res extensa*); and mind, whose essential property is that it thinks (*res cogitans*). Descartes' conceptual distinction between the material body and immaterial soul is similar in some respects to Plato's conceptual distinction between body and soul. The soul is only contingently and not necessarily related to the body and thus can exist independently of it (Descartes, 1641/2000, p. 6). Unlike Plato, Descartes does not divide the soul into parts. He also holds that mind and body can causally interact through the pineal gland in the brain The main difference between Plato's and Descartes' soul-body dualism is that Plato accepts, and Descartes rejects, the assumption that the soul animates the body.

The idea from Plato and Aristotle of the soul as an animating or organizing principle of the body has evolved into the empirical fact that the brainstem sustains circulation, respiration, and other functions necessary to keep a human body alive. The idea of intellect in Aristotle and Aquinas has evolved into the idea that the capacity for conscious thought and behaviour is generated and sustained by integrated functions of the brainstem, thalamocortical, and corticocortical networks. Yet while this conscious capacity is nested in neurobiology and neural mechanisms, it is not reducible or identical to and cannot be explained away by them (cf. Churchland, 1986).

Consistent with the relation between Aristotle's account of the intellect and the hylomorphic composite, the phenomenology and content of mental states are emergent features of integrated information in the brain that necessarily depend on but cannot be completely explained in neural terms (Mahner & Bunge, 1997, 202ff.). This can be described as non-reductive materialism (Baker, 2009). What it is like to be conscious, and the external states of affairs to which our thoughts are directed, resist complete neurobiological explanations because they are not located in the brain. The conscious mind is an emergent feature of a nested, multilevel, neural hierarchy. More complex regions of the brain interact with less complex regions in generating and sustaining awareness. Cortical regions (frontal-parietal) are more complex than subcortical regions (upper brainstem, thalamus) in mediating sensorimotor, cognitive, and affective processes (Feinberg & Mallett, 2018, pp. 71–72). "When higher, more complex levels are added to and interact with lower levels, the system as a whole acquires new or even novel (never-before-existing) features", specifically, consciousness (71). Brain and mind are two levels of a unified system of information processing and representation in an organism. This system cannot be explained by ontological or epistemological reductionism. It is not just a feature of a simpler or more basic kind. Nor is it just a function of its known component parts (Bennett & Hacker, 2022, pp. 415 ff.).

Among the conceptions of the soul that we have described, Aristotle's is the closest to our conception of the neurodynamic soul because it takes brain (matter) and mind (form, intellect) to be unified aspects of a human being (Charles, 2021; Goodman & Caramenico, 2013). *Psuche* is a more fundamental concept than *persona* because how we appear to and interact with others depends on neural and mental capacities. Our conception of the soul rejects historical and contemporary dualistic and mechanistic accounts of thought and behaviour. These include John Carew Eccles' claim that mind and brain are separable but can interact during activation of the cerebral cortex (Eccles, 1951) and the cognitive science view that the brain is a system of multilevel mechanisms (Craver, 2007; Bechtel, 2008). Our model is inspired by the neurodynamic theory of Walter Freeman. According to Freeman, the brain oscillates between information gathering and responses to the information. As a function of its perception-action cycle, integration, oscillation, and synchronisation of information about the world enable us to adapt to and engage with it (Freeman, 2000, 2001, 2003). These neural rhythms enable subjects to form and execute intentions necessary for this engagement. Neural and mental processes are 'entangled' in a representational stream that informs and guides subjects' adaptive behaviour (Buzsaki & Freeman, 2015; Freeman & Vitiello, 2016; Bressler et al., 2018).

In some respects, a precursor to Freeman's model of the intentional brain is Henri Bergson's conception of the brain "as an instrument of action" mediating motor and mental capacities (Bergson, 1912/2004, p. 83). Bergson rejects the idea that the neural underpinning of intentionality could be explained by localized centres in the brain and suggests that it is mediated by a distributed neural network (Trimble, 2016; 2020, p. 8). The critical role of motor capacities in agency shows that the neural rhythms mediating our actions are not just conscious but also unconscious. As we explain in Chap. 2, unconscious motor capacities are as critical as conscious motor and mental capacities in our interaction with the world. We cite examples of neuropsychiatric pathologies to support this claim.

Humans are psychophysical beings consisting of interacting and interdependent neural and mental processes. Mind and brain mutually influence each other in a series of re-entrant loops to maintain homeostasis within the organism and promote the subject's ability to respond and adapt to and engage with the world. Neural oscillation and synchronization in networks consisting of the ascending reticular activating system in the brain stem, projections from it to the thalamus, and projections from them to cortical regions generate and sustain the nature and content of our unconscious and conscious mental states (Mahner & Bunge, 1997; Tononi et al., 2016; Feinberg & Mallett, 2018). The thalamocortical and corticocortical networks generating and sustaining consciousness interact with the basal ganglia in enabling sensorimotor, cognitive, and affective functions. Conscious and unconscious mental states influence neural processes by ensuring that they function at optimal levels in forming accurate representations of the environment in which the subject lives and acts. Embodiment is as critical as embeddedness to selfhood and agency because somatosensory feedback between the body and brain, as well as proprioception, interoception, and exteroception are critical to how we conceive of ourselves and navigate the world (Gallagher, 2005). We develop this neurodynamic model and discuss its neuroscientific, psychological, social, and normative aspects in the next six chapters.

In Chap. 2, we explain how neurodynamics enables adaptive behaviour. Understood in neo-Aristotelian terms as a psychophysical process, the soul enables adaptability to the natural and social environment by providing us with the capacity for practical and moral reasoning and wisdom. This grounds excellence and flourishing in the realisation of our individual and collective interests. Within a neuropsychological framework, this involves a normative aspect of Ludwig Wittgenstein's epistemic notion of 'knowing how to go on' in the situations in which one finds oneself. It involves a broader understanding of normativity in Wittgenstein's discussion of rule-following in the *Philosophical Investigations* (1953, #224 ff.). This includes not only forming and executing individual action plans in navigating the world but also collective plans with others with whom we share it. The agent has the neurological and mental capacity and rational and moral confidence in *knowing how* she should go on in acting and interacting with others in a range of projects and *knowing that* she is justified in going on (Kripke, 1982; Ginsborg, 2020; Gallagher, 2020). This is consistent with Bergson's claim that the intentional brain enables us "to know how to use a thing" (Bergson, 1912/2004, p. 83). These are features of our neurodynamic conception of agency.

'Agency' is often used interchangeably with 'free will'. Some conceptions of free will define it in terms of whether causal determinism (the combination of natural laws and events in the past) precludes the ability to do otherwise. Hard incompatibilists argue that causal determinism precludes alternative possibilities of action and that we do not have free will. Libertarian incompatibilists argue that we do have the ability to do otherwise, that we have free will, and that causal determinism is false (Kane, 1996; Strawson, 2010, p. 10). Compatibilists argue that what matters regarding whether we have free will and can be morally responsible for our actions is not whether we have alternative possibilities of action but that we act without coercion, compulsion, or constraint (Dennett, 1984, 2003; Fischer, 1994). We act freely when we are capable of reflecting on our desires and deciding upon which of these desires we act (Frankfurt, 1971). We can be morally responsible for actions when we have the capacity to respond to reasons for or against performing them (Fischer & Ravizza, 1998). The critical mental events (beliefs, desires, intentions) in all these conceptions of free will precede actions. Flexibility and adaptability are not only about the present but also the future. They involve goal-directed behaviour and the ability to form and execute action plans. 'Agency' more accurately captures the future-oriented aspect of behaviour than 'free will'. Also, 'will' may be associated with the volitional component of executing these plans (O'Shaughnessy, 2008). This is too limited because motor, cognitive, and affective capacities are also necessary for persons to act (Spence, 2009). Accordingly, we use 'agency' rather than 'free will' in discussing neurobiological and normative aspects of human behaviour. Effective agency is not only the ability to translate intentions into actions but to do it in a way that considers future possibilities and responds to changing environmental circumstances. The contours of agency are more about the mind-brain relation and how events in the past influence it than about whether these events and natural laws determine it.

We develop the idea of knowing how to go on in terms of what continental philosophers have described as being-in-the world in Chap. 3 (Heidegger, 1927/1962, pp. 13ff.; Sartre, 1943/1958, 1945/2007; Merleau-Ponty, 1945/1962; Bergson, 1912/2004, p. 168ff)). Maurice Merleau-Ponty's emphasis on the body's engagement with the world is especially significant for the organism's or subject's adaptability to it. In the same vein, Bergson claims that "our intellect. .. is intended to secure the perfect fitting of our body to its environment" (Bergson, 1911/1988, p. ix; Trimble, 2020) The phenomenology and normativity of embodied purposeful action resists mechanistic or quasi-mathematical analysis. This type of action includes not only Western but also Third World, or Indigenous, perspectives of the world and the individual's place in it. The way in which the dynamic brain enables flexible and adaptive behaviour is particularly significant in allowing populations to survive radical changes in these environments or those to which they are displaced (Wexler, 2008, Chs. 4, 5). These changes are part of a social and cultural ethology that evolves in response to them. We describe the significance of work of the nineteenth-century neurologist and psychiatrist John Hughlings Jackson for this concept. Insights from his research into the human condition were scientifically grounded yet also shaped by an ethology of the social brain. This was based firmly in principles of neural association and neural networks in advanced mammalian brains (Hughlings Jackson, 1887; Franz & Gillett, 2011). We explain how the neurodynamics of Freeman and other researchers is a further development of Hughlings Jackson's work, how it has resulted in a better understanding of the relation between the brain and the world, and how this relation shapes human mentality and agency.

In Chap. 4, we further discuss how the dynamic brain enables flexible being in the world and promotes adaptability. We explain how neurodynamics transcends mechanistic and deterministic accounts of behaviour while remaining naturalistic. This involves a rapprochement of Anglo-American and Continental philosophy to understand the agent's place in the world. Neurophilosophy is not about a machine designed for specific operations involving information input and behavioural output but an open-ended relation of the brain to the environment in which fluid ways of perceiving, knowing, and acting emerge. These ways form an ethological niche of human genetic and epigenetic evolution. The effects of this relation on human agents are marked, particularly regarding discourse. This discursive aspect of behaviour becomes a living medium of human development that is not a kind of calculus with a rigid grammatical structure. Instead, it is part of a wider social and cultural framework. As Wittgenstein noted in his later work, natural language cannot be adequately explored and understood within an artificial structure of language stripped of how it is used (Wittgenstein, 1953, 1969, 1991). This use occurs within dynamic and evolving holistic incursions into the wider context of human life.

We elaborate the discursive aspect of being human in Chap. 5. As part of membership in mutually supportive groups, being discursive allows a level of communication about the natural and social environment that makes us capable of responding to its challenges (Harre & Gillett, 1994). Our capacity for discourse is one of the cognitive capacities that allows us to imagine, meet, and overcome these challenges in an epigenetic and collective manner. As a culturally developed capacity that is attuned to the world in which we live, discourse promotes adaptability. It allows for different meanings of our practices, depending on the contexts in which we use language. Being discursive is thus a key component of human agency. Our neurocognitively based discursive ability is realised in meaningful individual and collective conscious action. Discourse has a critical role in the unity of consciousness and unifying practical aspects of our lives. To illustrate this phenomenon, we describe the role of discourse in social integration and the narrative of a person's experience of and recovery from a brain injury. We provide examples of how brain injury and disease can adversely affect our experience of embodiment and cause us to lose our discursive capacity to interact with and communicate with others. This loss can impair adaptive and flexible being-in-the-world.

In Chap. 6, we discuss how consciousness allows us a wide discursive elaboration of being-in-the-world. A wide spoke of perception and interpretation, or 'seeing as', is available through imaginative constructions of possible experience. We have highly articulate techniques of sharing these constructions through grammar and the cognitive structures of thought and imagination. We have developed a language using terms of evaluation to assess techniques of interaction with the world and each other. They allow us to reflect on and control our lives using the full resources of a neurocognitive system consisting of perception, cognition, and action (Gillett, 1992). What Wittgenstein calls 'grammar' is an important part of this system and its function in social and cultural contexts. Broadly construed, grammar consist of a set of rules of a 'game' or other social activity and the moves the rules allow. Our understanding of these rules is manifest in how we apply them (Wittgenstein, 1953, #95ff.; 1991, I, #23ff). Intentionality is critical in this activity. It involves not only the ability to form action plans but also to anticipate how others will act and a willingness to cooperate with them based on trust in their decision-making. Intentionality and other cognitive capacities are components of our practical and moral reasoning of going on in and engaging with the world.

We discuss a broad normative concept of goodness in Chap. 7. Goodness is a natural disposition the exercise of which allows us to act in accord with our own interests and the interests of others and thereby flourish. Inspired by Philippa Foot's *Natural Goodness* (2001), our discussion shows the connections between excellence, practical wisdom, and flourishing. We explain the full scale of the normative aspects of 'going on' in acting and interacting with others discussed in Chap. 2. Some of the intentions we form and express in discourse, which is a type of action, relate us to others within the context of our shared adaptation to the world. Within this context, we have various ways to be good human beings both prudentially and morally. We do this in binding ourselves to each other through promises, agreements, jointly formulated strategies, and other arrangements to promote and enhance not only flexible activity but also human well-being in the present and future. A good life comprises a number of ways of conducting oneself that spring from a nuanced combination of different types of excellence, wisdom, and flourishing at theoretical and practical levels.

In the Epilogue, we summarise the main points from the seven chapters We emphasise the main aspects of neurodynamics and the idea of the soul as a psychophysical process consisting of interacting and interdependent neural and mental functions of embodied subjects acting and interacting with other subjects in natural, social, and cultural environments. We are capable of flexible and adaptive behaviour because of interactions between and the brain, mind, body, and world. We spell out some of the implications of the neurodynamic soul for philosophy of mind and rational and moral behaviour.

Neurodynamics, the later Wittgenstein, and indigenous perspectives on being-in-the-world may at first blush seem disparate topics lacking conceptual integration. But they are integrated in the sense that the idea of adaptability, and how the dynamic brain enables it, is an extension of Wittgenstein's discussion of knowing how to go on in the natural and social environment. 'Going on' is a way of describing how persons navigate and engage with the world. We do this through our use of language in conversation and other forms of behaviour. They are critical features of the brain and how it responds to and is shaped by what is external to it. Persons and their brains are not jus*t biological* organisms but also *social* organisms. The indigenous perspectives we discuss offer a different way of describing what it means to go on in and interact with the social and cultural milieu in which we are embedded. The role of indigenous languages, like the Māori language, in discursive behaviour and adaptability is a particularly important feature of these perspectives, which emphasise a communal sense of being-in-the-world They broaden the idea of human adaptation to the world. We develop this integrative model through Chaps. 2–6 and show how it generates a broadly normative conception of natural goodness in Chap. 7 and the Epilogue.

References

Aquinas, T. (1973–1980). *Summa Theologiae* (Fathers of the Enish Dominican Province, Trans.). Blackfriars.

Aristotle. (1984). The complete works of Aristotle, *Revised Oxford Translation*, Volumes One and Two, trans. and ed. J. Barnes. Princeton University Press.

Baker, L. R. (2009). Non-reductive materialism. In B. McLaughlin & A. Beckerman (Eds.), *The Oxford handbook of philosophy of mind* (pp. 109–120). Oxford University Press.

Bechtel, W. (2008). *Mental mechanisms: Philosophical perspectives on cognitive neuroscience*. Taylor & Francis.

Bennett, M. R., & Hacker, P. M. S. (2022). *Philosophical foundations of neuroscience*, 2nd ed. Wiley-Blackwell.

Bergson, H. (1911/1988). *Creative evolution* (A. Mitchell, Trans.). Dover Books.

Bergson, H. (1912/2004). *Matter and memory: Essay on the relation of body and spirit* (N. M. Paul & W. S. Palmer, Trans.). Dover Books.

Bressler, S., Kay, L., Kozma, R., Liljenstrom, H., & Vitiello, G. (2018). Freeman neurodynamics: The past 25 years. *Journal of Consciousness Studies, 25*, 15–32.

Broadie, S. (2001). Soul and body in Plato and Descartes. *Proceedings of the Aristotelian Society, 101*, 295–308.

Buzsaki, G., & Freeman, W. (2015). Editorial overview: Brain rhythms and dynamic coordination. *Current Opinion in Neurobiology, 31*, v–ix.

Charles, D. (2021). *The undivided self: Aristotle on the mind-body problem*. Oxford University Press.

Churchland, P. S. (1986). *Neurophilosophy: Toward a unified science of the mind-brain*. MIT Press.

Craver, C. (2007). *Explaining the brain: Mechanisms and the mosaic unity of neuroscience*. Oxford University Press.

Dennett, D. (1984). *Elbow room: The varieties of free will worth wanting*. MIT Press.

Dennett, D. (2003). *Freedom evolves*. Viking.

Descartes, R. (1641/2000). *Meditations on first philosophy* (J. Cottingham, Trans. and Ed.). Cambridge University Press.

Eccles, J. C. (1951). Hypotheses relating to the brain-mind problem. *Nature, 168*, 53–57.

Feinberg, T., & Mallett, J. (2018). *Consciousness demystified*. MIT Press.

Firth, J., Conlon, C., & Cox, T. (Eds.). (2020). *Oxford textbook of medicine* (6th ed.). Oxford University Press.

Fischer, J. M. (1994). *The metaphysics of free will: An essay on control*. Blackwell.

Fischer, J. M., & Ravizza, M. (1998). *Responsibility and control: A theory of moral responsibility*. Cambridge University Press.

Foot, P. (2001). *Natural goodness*. Oxford University Press.

Frankfurt, H. (1971). Freedom of the will and the concept of a person. *Journal of Philosophy, 68*, 5–20.

Franz, E., & Gillett, G. (2011). John Hughlings Jackson's evolutionary neurology: A unifying framework for cognitive neuroscience. *Brain, 134*, 3114–3120.

Freeman, W. J. (2000). *Neurodynamics: An exploration in mesoscopic brain dynamics*. Springer.

Freeman, W. J. (2001). *How brains make up their minds.* Cambridge University Press.
Freeman, W. J. (2003). Neurodynamic models of brain in psychiatry. *Neuropsychopharmacology, 28,* S54–S63.
Freeman, W., & Vitiello, G. (2016). Matter and mind are entangled in two streams of images guiding behaviour and informing subjects through awareness. *Mind & Matter, 14,* 7–24.
Gallagher, S. (2005). *How the body shapes the mind.* Oxford University Press.
Gallagher, S. (2020). *Action and interaction.* Oxford University Press.
Gillett, G. (1992). *Representation, meaning and thought.* Clarendon Press.
Ginsborg, H. (2020). Wittgenstein on going on. *Canadian Journal of Philosophy, 50,* 1–17.
Goodman, L., & Caramenico, D. G. (2013). *Coming to mind: The soul and its body.* University of Chicago Press.
Harre, R., & Gillett, G. (1994). *The discursive mind.* Sage Publications.
Heidegger, M. (1927/1962). *Being and time* (J. Macquarrie & E. Robinson, Trans.). Blackwell.
Hughlings Jackson, J. (1887). Remarks on the evolution and dissolution of the nervous system. *British Journal of Psychiatry, 33,* 25–48.
Kane, R. (1996). *The significance of free will.* Oxford University Press.
Karasmanis, V. (2006). Soul and body in Plato. *International Congress Series, 1286,* 1–6.
Kripke, S. (1982). *Wittgenstein on rules and private language.* Harvard University Press.
Mahner, M., & Bunge, M. (1997). *Foundations of biophilosophy.* Springer.
Massimini, M., & Tononi, G. (2018). *Sizing up consciousness: Towards an objective measure of the capacity for experience.* Oxford University Press.
McLaughlin, B., & Beckerman, A. (Eds.). (2009). *The Oxford handbook of philosophy of mind.* Oxford University Press.
Merleau-Ponty, M. (1945/1962). *Phenomenology of perception* (C. Smith, Trans.). Routledge.
O'Shaughnessy, B. (2008). *The will: A dual aspect theory* (2nd ed.). Cambridge University Press.
Plato. (1961). *Phaedo and Republic.* In *The collected dialogues of Plato,* trans and ed. E. Hamilton & H. Cairns. Princeton University Press.
Sartre, J.-P. (1943/1958). *Being and nothingness* (H. Barnes, Trans.). Methuen.
Sartre, J.-P. (1945/2007). *Existentialism is a humanism* (C. Macomber, Trans.). Yale University Press.
Spence, S. (2009). *The actor's brain: The cognitive neuroscience of free will.* Oxford University Press.
Strawson, G. (2010). *Freedom and belief* (revised edition). Clarendon Press.

Tononi, G., Boly, M., Gosseries, O., & Laureys, S. (2016). The neurology of consciousness: An overview. In S. Laureys, O. Gosseries, & G. Tononi (Eds.), *The neurology of consciousness: Cognitive neuroscience and neuropathology*. Elsevier.

Trimble, M. (2016). *The intentional brain: Motion, emotion and the development of modern neuropsychiatry*. The Johns Hopkins University Press.

Trimble, M. (2020) Phenomenology: A neuropsychiatric perspective. In W. Agrawal, R. Faruqui, & M. Bodani (Eds.), *Oxford textbook of neuropsychiatry* (16pp). https://doi.org/10.1093/med/9780198757139.001.0001.

Warren, J. (Ed.). (2009). *The Cambridge companion to Epicureanism*. Cambridge University Press.

Wexler, B. (2008). *Brain and culture: Neurobiology, ideology, and social change*. MIT Press.

Wittgenstein, L. (1953). *Philosophical investigations* (G. E. M. Anscombe, Trans.). Macmillan.

Wittgenstein, L. (1969). *On certainty* (D. Paul & G. E. M. Anscombe, Trans.). Blackwell.

Wittgenstein, L. (1991). *Philosophical grammar* (A. Kenney, Trans.; R. Rhees, Ed.). Blackwell.

Zeman, A. (2020). Mind and brain: Building bridges between neurology, psychiatry and psychology. In J. Firth, C. Conlon, & T. Cox (Eds.), *Oxford textbook of medicine* (6th ed.). Oxford University Press. https://doi.org/10.1093/med/9780198746690003.0568

CHAPTER 2

Neurodynamics and Adaptive Behaviour

If we trace the soul through its Greek sources to Aristotle, we are led to the concepts *psuche, arete, phronesis,* and *eudaimonia*. These concepts focus on the integrated mental, physical, and social life of persons and their species-typical attributes. They include cognition, emotion, and highly articulate communication. The full complement of these attributes forms the complexity of biopsychosocial life and its widespread and interwoven characteristics. Through them we come to see the joy of doing what is good, or at least what, for oneself as a human being among others is good. This normative claim depends on an account of agency involving relations between the brain, mind, and body of the organism, or subject, situated in the world.

Georg Northoff argues that the traditional philosophical mind-body problem is the "wrong path by which to tackle the question of the existence and reality of mental features" such as consciousness (Northoff, 2018, p. 1). He proposes that analysing the relation between the world and the brain offers a more satisfactory explanation of these features than the relation between the brain and the body. The brain is not an isolated organ but one that functions within the time and space of the world that surrounds and supports it (Northoff, 2021). This complex temporospatial interaction forms the ecology of the brain as an embodied organ that allows the organism to adapt to the world by mediating the relation between them. (Fuchs, 2018; 2021). "The brain is an enabler—an

G. Gillett, W. Glannon, *The Neurodynamic Soul*, New Directions in Philosophy and Cognitive Science, https://doi.org/10.1007/978-3-031-44951-2_2

instrument that brings us into contact with the world" (Zeman, 2008, p. 316). The mind is not the product of the brain alone but of the dynamics of the world in which the organism, or subject, is embedded. "It is not the brain, but the living human person as a whole who thinks, feels and acts" (Fuchs, 2018, p. 1). Our concept of the neurodynamic soul as a psychophysical process involves these same types of neural ethology.

"While neural activity is related to specific tasks or stimuli, specifically stimulus-induced or task-related activity, the brain shows intrinsic activity" (Northoff, 2018, p. 1). These types of activity are components of the brain's spatio-temporal structure. They are central to the world-brain relation and how it generates and sustains mentality, thereby relating us socio-politically to others. But what is the purpose or adaptive function of mental features and the neural activity that mediates them? Neural and mental activity as properties of human organisms sustain our organismic integrity. They enable flexible behaviour in allowing the subject to respond, adapt to, and engage with natural, social, and cultural environments—the world—in which they are embedded. In addition, neural and mental activity respond to these environments to enable us to act in accord with our personal and interpersonal interests and thereby flourish. Because human organisms, or subjects, are embodied, neural and mental functions are also influenced by the body proper and its role in interoception, exteroception, proprioception, and somatosensory processing. These processes enable the subject to experience herself as being and acting in the world, as an embodied and embedded subject (Gallagher, 2005, 2020). The mind of the human subject is therefore embodied, embedded, and enacted via human language and cognition, a pinnacle in organismic neuro-cognitive development (Rowlands, 2010, Ch. 3; Gallagher, 2020).

RHYTHMS IN THE BRAIN

Neural systems generate and sustain the mental systems that emerge from them when they reach a certain level of complexity. Mental function can influence neural function, as studies of the effects of cognitive behavioural therapy on the brain have shown (Goldapple et al., 2004; Fordham et al., 2021). Neural and mental systems are influenced by interaction between immune, endocrine, cardiovascular, and enteric systems and the brain. They are also influenced by how the organism or subject perceives and responds to the environment (prominently socio-political). Accordingly, the appropriate relation in explaining human thought and action is not mind-body, mind-brain, or brain-world but mind-brain-body-world. This holistic framework forms a series of re-entrant loops between these features of human life so that we must reject any reductionist explanation of the mind in entirely neural terms. It also provides a link between empirical, or descriptive, and normative aspects of the neurodynamic soul and its role in promoting responsiveness, adaptability, engagement, and flourishing in actions that realise our immediate and long-term individual and collective plans.

Intrinsic activity in the brain does not imply that it is an initial condition that precedes and is immune to external influence. The world-brain relation is not a disjunctive but a conjunctive relation. World and brain are not distinct entities but interacting and interdependent aspects of an organism or subject whose thought and behaviour occur within natural, social, and cultural environments. Gyorgy Buzsaki explains that "the neuronal 'signal' in response to a given environmental perturbation of the brain state is not an initial condition but, rather, a modification of a perpetually evolving network pattern in the brain's landscape" (2006, p. 276; 2019). Neural oscillation and synchronisation bring together the collective behaviour of neurons so that they have maximal impact on their targets. This depends on activity in a hierarchy of multiple re-entrant loops in cortical and subcortical regions. These loops suggest that the causal processes involved in this behaviour are circular rather than linear (Woodward, 2021, Ch. 2). Sensory input and motor and behavioural output occur in a neural landscape that is "under the supervision of the body, environment, and interactions with other brains" (Buzsaki, 2006, p. 31). The last are, we should always remind ourselves, interactions between embodied beings.

In his later writing, Wittgenstein discusses and provides many examples of how meaning and understanding are human forms of life. Their complexity is not reducible to simplistic or causally mechanistic concepts. In Wittgenstein's terms, if one is living well as a thriving human creature that is because one knows how to conduct oneself in a situation of the kind in which one finds oneself and therefore can go on in an assured and satisfying way. *Eudaimonia*, on this account, becomes conceptually related to *phronesis* in that feeling satisfied or fulfilled is the result of knowing when one knows how to go on. A person displays *arete*, as virtue or excellence, when one masters this technique and others in natural and social settings. The more generally fulfilled and competent one is and the less one excuses and corrects oneself, the better life is for any human being. The characteristic function of *psuche* enabling reasoning, decision-making, and the completion of action plans, connects *phronesis, arete,* and *eudaimonia.*

It is mistaken to think of human behaviour as solely driven by merely appetitive rewards such as those applied in early operant conditioning. An animal, it turns out, is guided far more by the opportunity to be in stimulus conditions associated with a frequent behavioural activity rather than a primary reinforcer per se. That indeed makes a lot of sense as any creature should develop and wish to develop a repertoire of species-typical responses however those became part of its behavioural repertoire (for instance by imitation or socialisation). Thus, within a neuroscientific framework, the soul (psyche) has evolved from being a relatively simple into a dynamic and complex psychophysical and neurodynamic process interacting with the world through prediction and self-correction or self-formation (autopoiesis) on the basis of an evolving adaptive balance and ethological success.

One may feel 'at odds' in an unfamiliar or *unheimlich* setting to which one has never attuned oneself. That is to be expected but may initially strike one as a prima facie paradox in operant conditioning. Animals work to get access to conditions in which they have learned to work harder. But if we change the description to 'make a high frequency response' and then realise that this phenomenon is progressive and flexible, adjusting itself to unexpected contingencies as they arise, it begins to make increasing sense. We develop 'ways of going on' in each life context. Thus, the soul has evolved from antiquity, progressing from a relatively simple to a complex psychophysical process maximising organismic outcomes in any situation. For human beings, this is hugely enhanced by our communication and symbolic activities including religion and a sense of the good beyond simple rewards. At a certain point, we therefore move beyond the neurodynamics of the animal kingdom and into the domain of 'the noumenal', or the conscious experience of phenomena with many layers of meaning.

One alternative possibility whereby an animal would become disorganised and unwell could arise if it was exposed to a painful punishment regime, especially if it was brain-damaged (Gillett & Webster, 1975, 14.6.883). In one paper, consistent with its time, Grant Gillett and D.M. Webster located the failure of conditioning in a basal ganglia lesion, but now it seems that a better conclusion would be that the regime itself was behaviourally disorganising and that the neural disturbance of function merely exacerbated that pathological tendency. The poor rats were functioning like psychologically disturbed creatures that were trapped into a malignant reward and punishment schedule where the contingencies were confusing and unstable such that there was no reliable high-frequency pattern of responding that would allow them to function well. All creatures adapt by going on in a certain way that is sustainable or successful in a certain ecological niche. The conceptual structure on which Wittgenstein eventually built his philosophy of mind involves this natural phenomenon with its 'way of going on'. We need some work to get from adaptation and animal behaviour to language and conceptual or neuro-cognitive being, but the work is, in itself, illuminating. It is particularly illuminating when we consider the behaviour of disturbed young people, especially those from cultural minorities who are marginalised or alienated within society. The rhythms of the brain in such a young person are cross-grained in such a way that harmonious living with others is almost impos-sible as, like the rats in the experiment we cited, the contingencies disrupt the formation of stable patterns of neural function.

Avoiding a dualistic view whereby where mind is conceptually distinct from body and brain, a nuanced naturalism about human affairs and human good does not seek to distort the interwoven complexities of life in the world. These are more complex and ramified in the human case than for any other creature, adapted as they are to both the real and the imaginary contexts of human action. The real contexts are relatively straightforward to conceptualise in continuity with animal adaptation. But the imaginary contexts force us to consider neural rhythms which do not easily fit those more straightforwardly biological dynamics.

The range of human cognition and its extension into counterfactual and uniquely configured situations requires some mechanism (or rather technique) which allows for distinctions only established by adopting some objective point of view. Such finely characterised relations to what we call 'truth' introduces us not only to qualitatively distinguishable situ-ations or states of affairs but also to the difficult notion of 'singular thought'. It is no wonder that this highly nuanced view occupied the attention of realistic philosophers of meaning and language such as Gareth Evans and John McDowell (Evans, 1982; McDowell, 1998). Such an approach to human cognition is 'nuanced' in the sense that it involves mind and brain as interacting aspects of a neurodynamic system active in human organisms interacting with the world and sharing highly detailed information about that world through language. We discuss this idea in more detail in Chap. 5 (Being Discursive).

A philosophical psychology of this detailed form, informed as it is by the philosophy of language and thought, offers us a unique insight into the rhythms of the holistic and cognitively sophisticated human mind. In so doing, it explains how human beings modify the world in a two-way exchange so that the conditions of adaptation are not fixed but continually, and to some extent predictably, changing. This is adaptive resonance theory (ART) (Grossberg, 2013). Human adaptation, therefore, to some extent, incorporates the results of Aristotle's *phronesis* as a set of well-practised techniques or skilled life performances which one deploys in the situation where one has mastered such skills and brings them into action in response to certain cognitively open conditions. Such informed skills make up the rhythms of everyday life when a person is functioning normally, and their net effect is to produce an adaptive repertoire of responses to various challenges that arise within one's ethological niche. In these respects, our adaptation comprises many instances of what Wittgenstein describes as "mastery of a technique".

This conceptual path avoids both reductive naturalism and its embedded mechanistic approach to human and biological life and is therefore important in developing a sound basis for the neuroscientific foundation of ethical behaviour. Appeal to the brain alone cannot explain this behaviour. Rather, certain neural structures and processes enable the mental capacities through which we recognize and respond to the demands of the environment. They also enable us to respond to the needs, interests, and rights of others. The Aristotelianism involved is truly neurocognitive and sensitively nuanced in favour of a philosophical exploration of a human being as an epigenetically self-modifying creature in a dynamic environment that shapes and is shaped by his or her own activity.

AGENCY AND ADAPTABILITY

The relations between brain, mind, body and world form a neurodynamic conception of agency. We are effective agents capable of flexible behaviour when we can mentally simulate counterfactual states of affairs, anticipate future challenges, and form and execute action plans in light of them. These capacities enable adaptability to these situations and others not yet encountered but possibly real and pressing in the future. A neurodynamic model conceives of agency as a continuous neural and mental process shaped by the organism's interaction with its surroundings and its rhythms of response to them. These rhythms cannot be unpacked like operations in an industrial system or digitalized like a computer program. Our adaptive ways of going on cannot be described entirely in general neurocognitive terms because they involve personal styles, techniques, or characteristic modes of sensorimotor, cognitive, affective, and volitional engagement with the physical and social environment.

Freeman has shown the dynamic brain in action in several carefully recorded neuroscientific experiments on living normally functioning animals. He demonstrated that the brain oscillates between an information-gathering mode and a practiced routine of response (Freeman, 2000, 2001; Buzsaki & Draguhn 2004; Buzsaki & Freeman, 2015). The discovery was originally made by examining the processing activity of live rabbit brains, but the relevance of Freeman's neurodynamic model is much wider (Bressler et al., 2018). It tracks the brain from its engagement with the environment, through acquiring and processing information, to the development of adaptive behaviours in human subjects. One significant implication of the model is its perception-action cycle. This includes neural oscillations and synchronization generating the capacity for intentionality (action planning) and decision-making (executing plans) necessary to engage in adaptive behaviours (Bratman, 1987; Fuster, 2013, pp. 87ff.; Freeman & Changeux, 2015; Liljenstrom, 2018; 2022). This dynamic process shows us to be effective agents in the world as the relevant processes maintain an equilibrium in the brain allowing neurocognitive control of our thought and action suited to a life context. At the mental level, they allow us to respond appropriately to reasons for or against different actions. That ability allows us to control which actions we perform and take responsibility for them by matching them to life plans (Fischer & Ravizza, 1998; Bratman, 2007).

The sensorimotor system develops gradually over the lifetime of an organism, becoming attuned to an ecological niche. At the species level, it develops cumulatively in evolutionary time as we undergo genetic and environmental change. The neural rhythms in this system change in timescales from milliseconds to millennia. This results in the emergence of a wide repertoire of responses to different circumstances. Each response arises from a core of component responses with genetic and epigenetic elements, yielding an increasingly complex and adapted ethological repertoire of responses to environmental contingencies. An example of adaptability in the organism is how the eye pupil adjusts to momentary perceived illusory movement as a compensatory mechanism to anticipate what it will perceive in the next moment (El Haj & Lenoble, 2018). This is just one aspect among many showing how the brain is constantly trying to anticipate the future and allows us to do something similar at the mental level in goal-directed behaviour (Laeng et al., 2022). Each organism or agent has a distinct set of developmental and agentive features as it faces a unique set of changing external circumstances. Individual acts by individual members of the species follow a probabilistic rather than deterministic pattern and thus are not predictable (cf. Haynes, 2010; Hohwy, 2013). This is a function of epigenetic changes in neural and mental processing driving human behaviour.

Bruce Wexler explains some of the cultural aspects of epigenetic influences on the brain. It is the "ability to shape the environment that in turn shapes our brains and has allowed human adaptability and capability to develop at a much faster rate than is possible through alteration of the genetic code itself" (pp. 2–3). From youth to adulthood, "the shaping of brain function through culture also means that processes that govern the evolution of societies and culture have a great influence on how our individual brains and minds work" (p. 3). Wexler explains changes in the relationship between the individual and their environment in the transition from late adolescence to early adulthood.

> During the first part of life, the brain and mind are highly plastic, require sensory input to grow and develop, and shape themselves to the major recurring features of their environments. During these years, individuals have little ability to act on or alter the environment but are easily altered by it. By early adulthood, the mind and brain have elaborately developed structures and a diminished ability to change those structures. The individual is now able to act on and alter the environment, and much of that activity is devoted to making the environment conform to the established structures. In both periods, however, there is a neurobiological imperative that an individual's internal neuropsychological structures match key features of his or her external environment, a principle of internal-external consonance'. (Wexler, 2008, p. 5)

The motor effects of responding to external events form a set of skills. These skills involve unified brain-systems and mental-systems levels of the organism and subject. As noted in the Introduction, both systems are entangled as representational streams that inform and guide action (Buzsaki & Freeman, 2015; Freeman & Vitiello, 2016; Bressler et al., 2018). This entanglement allows adaptive motor and cognitive responses to these events cumulatively structuring a human life. Freeman's neurodynamic model describes the intentional brain as an evolving structure and process that evolves from the inherited behavioural patterns of the animal species to fashion novel responses to meet the sensorimotor and cognitive demands of the immediate environment that interacting members of the species are adapted to (Dennett, 2003).

MULTIFACETED AGENCY

Twentieth-century Russian neuropsychologist Alexander Luria emphasised the multifaceted aspects of the environment and its influence on neural and mental function (Luria, 1973). Indeed, he explained the brain-mind-world relation long before neurophenomenological theories of the embodied, embedded, and enacted brain and mind. For Luria, the environment that shapes the brain and mind is not just physical or natural but also cultural and historical. Cognition is a culturally constituted phenomenon. Rejecting localised and modular concepts of the brain. Luria theorised that the brain is part of a functional system that extends into the cultural-historical world. This extracortical organization is dynamically intertwined with the neural organization of the working brain. It is Luria's great legacy that has significantly shaped neuropsychology in the twenty-first century (Christensen et al., 2009).

This integrated model of the neural, mental, and environmental bases of action provides a more plausible explanation of behaviour than an agent-causal model focusing on the agent as the origin or ultimate source of action unaffected by the effects of natural laws and events in the past (Kane, 1996; Clarke, 2003; Strawson, 2010). It is also more plausible than a mechanistic model focusing on internal neural parts and processes that realise organism-level functions in a bottom-up way (Craver, 2007; Bechtel, 2008). These models ignore or fail to pay adequate attention to the brain-mind-environment interaction necessary for flexible behaviour The organism's—the agent's—brain is involved in this activity as a unified site of inclusive information processing and response. It cannot be dissected into identifiable neural or mental compartments, each with its own causal and measurable triggering conditions. Miguel Nicolelis points out that "the brain is not a mechanism; it's an organism. And organisms are not built. They evolve according to evolutionary processes and events that cannot be reasonably predicted" (cited by Miles, 2015). One might quibble and insist that the brain is part of a living and ethologically situated organism, but the point remains cogent.

The source of our actions is not an increase in activity in specific regions of the brain at a specific time. Indeed, 'source' is misleading because it suggests a single causal factor in what is instead a holistic process consisting of multiple factors that influence our actions. These include multiple connected neural circuits and how their functions are influenced by the environment. Neural rhythms enable agency by responding to factors inside and outside the brain. As Gyorgy Buzsaki states: "The neuronal 'signal' in response to a given environmental perturbation of the brain state is not an initial condition but part of an ever-changing pattern in the brain's landscape" (Buzsaki, 2006, p. 276). Consistent with our comments about the epigenetic expression of the genetic basis of behaviour, Buzsaki states that "the main pathways [of brain and behaviour] are genetically determined, but the fine-tuning of connections ('calibration' by the output-input match) is under the supervision of the body, environment, and interactions with other brains" (2006, p. 31). Brain-brain interaction is one component of interaction between human organisms within the natural, social, and cultural milieu (Buzsaki, 2019, pp. 53ff.).

In more recent work, Buzsaki argues that "the brain is a self-organized system with preexisting connectivity and dynamics whose main job is to generate actions and to examine and predict the consequences of those actions". He refers to this as an "inside-out" strategy. This is in contrast to an "outside-in" strategy, where the brain's "task is to perceive and represent the world, process information, and decide how to respond to it" (2019, p. xiii, Chs. 3, 5, 13). We adopt a view of agency that involves both of these strategies. The brain with its internal rhythms of oscillation and synchronization enables us respond and adapt to the world. But this ability depends on how we perceive information from the world. The normativity of thought and agency in assessing the rationality and morality of our choices is not a property of the brain but people whose brains mediate their relation to the natural and social environment (Bratman, 2014).

In response to an inclusive mode of information gathering from the environment, the organism develops new patterns of activity. These are based on established patterns. The more familiar and settled the environment, the more characteristic and familiar the organism's responses will be. But the organism always has the capacity for incremental innovation in novel environments. Constantly evolving neural rhythms that respond appropriately to these changes enable the organism to anticipate contingencies and meet current and future sensorimotor and cognitive demands of its milieu. These rhythms enable behaviour control in the neurocognitive realization of agency (Gillett, 2015). Freedom and creativity in choosing between different courses of action is a function of the brain and its reciprocal relation to the environment (Fuster, 2013, pp. 77ff.). These capacities are enabled by episodic memory (Schacter & Addis, 2007; Schacter et al., 2012) allied to lifelong behavioural plasticity. The ability to recall information about the past enables us to simulate possible, or counterfactual possibilities and courses of action and project ourselves into the future. This type of memory is thus critical for goal-directed behaviour. Adaptability implicit in our model of agency includes the ability not only to form and execute actions plans but also the ability to change plans in response to changing circumstances and execute them across a range of these circumstances. There is no algorithm that could predict which actions among these counterfactual possibilities we would perform because they are influenced by unpredictable environmental contingencies.

Possible worlds was a frequently discussed topic in metaphysics in the 1970s and 1980s. These are abstract or notional ways in which the actual world might have been (Loux, 1980, Chs. 9, 12). Much of the discussion revolved around possible-world semantics, where a world was defined as a maximally consistent set of propositions. Different authors devised theories to establish truth-conditions for claims about these worlds. Our discussion of counterfactual or simulated possibilities as the bases of our action plans shows that possible worlds are more relevant to agency than to the philosophy of language. They are not complete but not-yet-realized states of affairs reflecting how individuals perceive possible courses of action in projecting themselves into the future and choose some courses rather than others. Agentic possible worlds are not sets of propositions but distinct sets of choices, actions, and their consequences. They are like the 'Garden of Forking Paths' in Jorge Luis Borges' short story of the same title, or Krzysztof Kieslowski's 1987 film, *Blind Chance*, where different choices in three distinct storylines can influence the rest of the main character's life. Our examination of discursive issues in Chap. 5 involves the philosophy of language to a considerable extent. But it pertains to how we go on in the world rather than to the truth or falsity of claims about the world, thought of extensionally.

The ability to simulate possible or counterfactual courses of action, and to form and revise action plans in light of this simulation, is related to the prediction-error theory. This involves reward learning in goal-directed behaviour. How we learn and adapt to situations depends on how we respond when our expectations fail to conform to reality and cannot be met. Dopamine neurons signal a prediction error between an expected future reward and whether or not one gains it (Schultz et al., 1997, 2017; Nasser et al., 2017; Sinclair et al., 2021). When one engages in temporal discounting about the future and opts for long-term over short-term reward, there is an expectation of reward in response to certain cues. It cannot be predicted whether one will gain an expected reward, though, because external events can thwart the realisation of the expectation. This unpredictability about long-range goals is part and parcel of rational decision-making. Reward learning is part of adaptive learning. But the key to adaptive behaviour is not so much the realisation or defeat of an expected reward but how one responds to success or failure in gaining or failing to gain it. This depends not only on cognitive processing in selecting goals but also on motor, affective, and volitional processing that allows one to achieve these goals. It also depends on whether or to what extent the natural and social environment allow the agent to adapt to it. Some external conditions may not allow an agent to adapt to or change them. They may be beyond the agents' creative control, their resilience, or their current response repertoires and therefore limit their agency. In some cases, the problem may lie not in the world but in the brain. Disruption of dopamine signalling in neuropathologies, as in Parkinson's disease (PD), can impair the mental capacities necessary for reward and adaptive learning and this impair a critical component of agency.

The idea of resilience as a mental capacity suggests that neural and environmental events do not completely determine adaptability. Some PD patients can use psychological techniques to 'will' themselves to perform certain movements despite dyskinesias (McNamara, 2011). Moreover, people with epilepsy can use different psychological techniques to avert a tonic-clonic seizure at the onset of the aura that precedes it (Zeman, 2004). A review of psychological management of the impact of epilepsy on patients revealed that psychological techniques had a significant effect, including a reduction in seizure frequency (Spector et al., 1999).

Perhaps the most notable example of will as resilience and determination is that of Lyova Zazetsky. During the 1941 Battle of Smolensk in WWII, Zazetsky sustained a severe head injury that caused extensive damage to the left occipital-parietal region of his brain. This resulted in fragmentation of his memory, visual field, and bodily perception. Despite these impairments, with the help of the neuropsychologist Luria, he kept a journal for years recording his thoughts and memories on a daily basis

(Luria, 1987). With excruciating effort, he wrote thousands of pages. Luria notes that Zazetsky "fought with the tenacity of the damned to recover the use of his shattered brain" (Luria, 1987, p. xx). Zazetsky's determination in trying to reshape his identity and construct a meaningful narrative after his brain injury shows how one can retain and exercise the will and thus the volitional component of agency, despite significant neurological impairment. It is an example of the human capacity to imagine, persist with, and realise possibilities of action despite being constrained by a damaged brain. (See also Mele, 2020).

This neurodynamic model aligns with Buzsaki's model of a hierarchy of multiple parallel loops in the brain. These loops involve excitatory pathways between the thalamus and neocortex and inhibitory pathways between the basal ganglia, pyramidal system, and cerebellum. Balanced excitatory and inhibitory neural processes, and the mental processes that emerge from them, promote adaptability and effective navigation of the world.

Ordinarily, highly connected and fast-acting conscious planning and executive functions and unconscious motor functions of behaviour coordination in practiced routines work together in a complementary manner. They may be dissociated, however, in individuals with basal ganglia lesions from stroke or traumatic brain injury. Although this damage to this subcortical region does not affect the cognitive contents of consciousness, it can have a profound effect on motor coordination and patterning that are essential to human agency. Damage and dysfunction in either cortical or subcortical networks mediating sensorimotor functions can impair the ability to translate intentions into actions identified with voluntary bodily movements. Recovery can be long and frequently punctuated by affective factors, as in the experience of one of the authors of the present work.

When motor functions are intact, dysregulation in pathways between the prefrontal cortex and anterior cingulate cortex regulating executive functions can cause avolition and the inability to will oneself to act (Spence, 2009; Hallett, 2007, 2016). Dysregulation of the nucleus accumbens can disrupt connectivity between motor and limbic areas and cause anhedonia, impairing the ability to form desires to act and impairing motivation. Cognitive, affective, and motor impairment can weaken the ability to initiate movement (Spence, 2009). Both avolition and anhedonia can thereby impair conscious planning and executive functions and thus can impair agency. These oft-encountered examples show that unconscious motor functions and conscious cognitive, affective, and volitional functions are necessary to form and complete action plans.

Neural Events, Time, and the Meaning of Actions

Neural, mental, and social factors are necessary to explain how we reason, decide, and act. Some neuroscientists have challenged the view that conscious intentions and decisions have a causal role in our actions. In much-discussed experiments conducted in the 1980s involving subjects who were asked to flex their wrist or finger, Benjamin Libet noted that unconscious brain events preceded the conscious intention to flex by 300–500 milliseconds. Readiness potentials associated with these events in cortical regions were recorded by electroencephalography (EEG) (Libet, 1985). Libet concluded that "the discovery that the brain unconsciously initiates the volitional process well before the person becomes aware of an intention or wish to act voluntarily ... clearly has a profound impact on how we view the nature of free will (Libet, 2004, p. 8). For many neuroscientists, the impact has been negative. Based on Libet's experiments, Patrick Haggard claims that because "conscious intention occurs after the onset of preparatory brain activity, [it] cannot ... cause our actions" (Haggard, 2008, p. 941).

Alfred Mele has argued that the timing of conscious intention in these experiments does not prove that it has no causal role in the outcome. Unconscious cerebral initiative alone does not provide a complete causal explanation of our decisions and actions. In addition to a proximal intention to act at a specific time, a distal intention at an earlier time can influence the content of and motivation to form and execute a proximal intention at the time of action (Mele, 2009, pp. 45ff.; 2020). Sensorimotor functions associated with readiness potentials may precede a conscious intention to perform a specific action at a specific time. But these functions are shaped by diachronic conscious cognitive, affective, and volitional functions that enable planning and goal-directed behaviour over extended periods of time. Explaining an action requires a broader temporal framework that includes but is not limited or neurally proximate to the time at which the act occurred; precursors integral to act-identity may extend much earlier than the physical action itself. These events, and the agent's response to them, can influence neural oscillation and synchronisation in brain regions mediating the sensorimotor and psychomotor processing underlying intentional behaviour.

There are additional limitations in Libet's experiments and their implications for free will. First, instructed movements in a laboratory setting are
not accurate reflections of how we act in the face of contingencies in the
real world and translate intentions into actions in our response to them.
Second, the experiments do not capture the phenomenology or perception of acting without coercion, compulsion, or constraint which often
accompanies our conscious intentional actions. This perception relies on
certain cortical and subcortical features of a person's brain (Hallett, 2016).
It is not reducible to the brain, however, but instead is an emergent feature of it as the brain responds to the demands of the environment, including its socio-political complexities.

Apparently shifting his focus to a broader social and cultural context of
action, Haggard notes that how a person makes decisions in response to
factors external to the brain can influence neural processes. These are part
of the normative aspect of knowing how to go on in different
circumstances.

> Interestingly, both decisions (to act or not act) have a strong normative ele-
> ment; although a person's brain decides the actions that they carry out,
> cultural factors and education teach people what are acceptable reasons for
> action, what are not, and when a final predictive check should recommend
> withholding action. Culture and education therefore represent powerful
> learning signals for the brain's cognitive-motor circuits. (Haggard,
> 2008, p. 944)

Emphasizing that human actions have a complex set of causes, Haggard
states that these causes "reflect the flexibility and complexity of our
response to our environment" (Haggard, 2011, p. 23). The causes of
most human actions are not limited to readiness potentials in the brain
measured at specific times but also include diachronic planning and delib-
eration shaped by the context in which the person lives, thinks, and acts.
Haggard's comments are consistent with Buzsaki's explanation of brain
function and how it is influenced by a person's interaction with the environment. They are also consistent with a broad interpretation of Freeman's
neurodynamic model. Neural oscillations and synchronization generate
and sustain intention and action-perception in decision-making through
the brain's response to factors external to it. One decides and acts not just
from neural events and processes alone but also because of the agent's
relationship with the world.

The context of action determines its meaning for the agent and those who hold them to account for what they do or fail to do. Sean Spence uses the following example to make this point. The Nazi salute of Hitler in his speeches to the German masses in the 1930s and 1940s and the Black Power salute of Tommie Smith and John Carlos at the 1968 Summer Olympic Games in Mexico City involved the same nerve fibres and the same neural events in motor and other regions of the cerebral cortex. But their actions had very different meanings for them because of the different times, places, and social and historical contexts in which they performed them (Spence, 2009, p. 395). The actions were more than just bodily movements resulting from readiness potentials or other events in the brain. "It is not the instant of the act but its context that seems to matter" (p. 395). The different contextual factors behind the meanings of these actions suggest that they are not initiated by particular neural events at discrete times but by complex distal intentions shaped by factors external to the brain extending into the past.

Neuropathologies and Agency

Earlier, we noted how stroke and traumatic brain injury can impair sensorimotor functions that ordinarily enable us to perform voluntary bodily movements. Different neuropsychiatric disorders can also impair not only the sensorimotor component of agency but cognitive, affective, and volitional components as well. It is instructive to describe and discuss some clinical examples of neuropathology to show how agency can be impaired by dysfunction at both neural and mental levels. This impairment has normative implications because it may support claims of mitigation or excuse regarding moral and criminal responsibility for one's actions. Cognitive science models of the mind typically cite psychogenic disorders to motivate these claims (Gennaro, 2015), some which reveal that this is yet another shortcoming of these models. Disruption in neural pathways mediating the mental capacities for behaviour control often also involve disruption in pathways between bodily systems (endocrine, immune) and the brain and the individual's relationship to the natural and social environment. They show how moral judgments about maladaptive behaviour associated with neuropathologies are shaped not just by accounts of the mind alone or mind-brain interaction alone but by adverse interaction between mind, brain, body, and environment.

This is holistic neuropsychiatry. Brain abnormalities impairing or undermining behaviour control can shed light on how different pathways enable such control in normal, healthy brains (Northoff, 2016). Philosophical accounts of agency as an entirely mental process consisting of desires, beliefs, intentions, and decisions are unsatisfactory because they ignore the neural substrate that mediates mental states as well as the body and the world that shape their content (Velleman, 1992; Davidson, 2001).

Sensorimotor impairment associated with degeneration of dopaminergic pathways in the substantia nigra of the basal ganglia can prevent patients with Parkinson's disease from performing voluntary bodily movements. Involuntary movements in dyskinesia induced by dopamine agonists to treat this disease involve different forms of motor impairment. The cerebellum regulates coordinated bodily movements. A stroke to this region of the brain can result in ataxia, the inability to perform smoothly coordinated movements. In healthy brains, these are actions we commonly perform without conscious effort. Moreover, psychomotor impairment in catatonia can also impair or, in severe cases, prevent, movement. While catatonia had been classified as a symptom of schizophrenia, it is now classified as a movement disorder (Castle & Buckley, 2015; Heckers & Walther, 2023). These are just some conditions indicating overlap between psychiatric and neurological disorders and reasons for categorising them not as distinct but as a single category of neuropsychiatric disorders (Zeman, 2014). Indeed, one of the authors (Gillett) encountered a disturbed Afro-Caribbean young man in London who was initially misdiagnosed as suffering a brainstem cardiovascular lesion but then recognised as having catatonic schizophrenia so that an appropriate psychiatric referral was made.

Disrupted connections between the basal ganglia and motor cortex can impede or preclude movement in people affected by movement disorders. Disrupted connectivity between limbic and cortical regions can downregulate the nucleus accumbens, the anterior cingulate cortex, and their projections to prefrontal executive regions. This can result in anhedonia and loss of the motivation to act, or avolition and loss of the ability to form and complete action plans. Cognitive and affective capacities interact. A person with anhedonia from depression may lack the motivation to plan or initiate actions. Motor and volitional processes also interact. An individual with a movement disorder like Parkinson's disease (PD) or apraxia may want to perform a certain bodily movement but not be able to do this because of motor impairment. These neuropathologies can cause mental paralysis in people affected by them in major depression and the negative subtype of schizophrenia (Castle & Buckley, 2015). Manic episodes in bipolar disorder I and hypomanic episodes in bipolar disorder II are both associated with an imbalance between limbic and prefrontal pathways. The resultant disrupted dopaminergic signalling can impair the ability to respond to reasons and refrain from acting in ways that harm oneself or others (Strakowksi, 2014).

In Parkinson's disease, the motor impairment resulting from dopaminergic degeneration can prevent or impair control of bodily movements and in this respect can limit an affected person's agency. PD may also impair the cognitive component of agency by disrupting connectivity between the basal ganglia and executive regions in the prefrontal cortex that ordinarily regulate working memory, reasoning, and decision-making (McNamara, 2011, pp. 69ff.). There are motor, limbic and associative circuits in the basal ganglia where the dopaminergic degeneration of PD occurs. The parafascicular thalamus (PF) is also implicated in this disease (Zhang et al., 2022). It projects to three parts of the basal ganglia: the substantia nigra, the caudate putamen, and the nucleus accumbens.

PD involves an imbalance of excitatory and inhibitory pathways involving the basal ganglia, thalamus, and motor cortex regulating motor output. It is an example of how neural dysregulation and degeneration in overlapping brain circuits can result not only in a movement disorder but cognitive and mood disorders such as anxiety and depression as well. Cognitive and motor impairment in PD can interfere with the ability to initiate and complete bodily movements. It can also interfere with the ability to perform complex long-range actions requiring combined motor and cognitive functions. Although it is typically characterised as a movement disorder, Parkinson's disease is also a cognitive, affective, and volitional disorder that can impair agency as much as if not more so than any other neuropathology (Powell et al., 2020). Depletion of dopamine in the basal ganglia in PD can also impair agency by impairing the capacity to update expectations about the future and modify behaviour. A dopamine deficit can thus interfere with adaptability. These considerations have implications for prediction-error theory and effective agency. In PD, as well as MDD and schizophrenia, neural dysfunction can interfere with the ability to use information about the world to predict events and form action plans in light of it. By impairing the neural basis of the mental capacities, neuropsychiatric disorders can impair adaptive learning. Rather than promoting a person's ability to engage with the world, they can contribute to disengagement from it.

Many of these pathologies are not just brain disorders but multi-system disorders and commonly involve disrupted neuro-endocrine and neuro-immune interaction. Chronic psychosocial stress can induce release of high levels of circulating hormones such as norepinephrine from the adrenal medulla and cortisol from the adrenal cortex. This can disrupt the activity of the hypothalamic-pituitary-adrenal axis and result in an adverse biochemical and physiological environment in the brain. It can disrupt homeostasis in the body and generate a chronic fight-or-flight or freeze response to benign environmental stimuli that are perceived as threats to the organism. These processes can impair the organism's ability to adapt to the environment. In a different process, release of proinflammatory cytokines in response to infection in the body or brain can disrupt neural transmitters mediating cognition and mood and contribute to or exacerbate major depression and generalized anxiety (Beurel et al., 2020). Yet another body-brain-mind disorder is when a chronic viral infection in the gut depletes serotonin, which in turn impairs memory and other cognitive functions (Wong et al., 2023). These psychiatric disorders can disrupt an affected person's perception of and response to the world, often involving withdrawal from the world or freezing behaviour in reaction to benign stimuli misperceived as aversive. They can cause mental paralysis by disrupting cognitive, affective, and volitional aspects of agency. As part of their disruptive effect on agency, these disorders can disrupt the experience of persisting through time as the same individual. They can therefore be described as neuropathologies of the self (Feinberg, 2011).

A neurophenomenological approach to brain-mind-body-world interaction can illuminate how neuropathologies can adversely affect a person's thought and behaviour and alter the relation of the self to the world. Northoff states that this "approach aims to link the phenomenology of psychopathology with underlying neural processes" (2015, p. 86). There is disruption of normal mental processing and how one experiences the world. Northoff focuses on depression and schizophrenia as two disorders in which the self is no longer able to adapt to the world because they have become alienated or isolated from it. He associates these maladaptive behaviours with abnormal resting state activity in cortical midline structures (2014, pp. 391ff., 2016, pp. 105ff.). The activity may be either hypoactive or hyperactive. These are the main neural correlates of consciousness and the phenomenology of the self. In depression and schizophrenia, dysregulation occurs mainly in cortical-limbic and cortical-striatal pathways (a subcortical-cortical-paralimbic network, p. 95). "In depression, there is increased directedness toward the own self and the body ('increased self- and body-focus'), while the directedness toward the environment is decreased ('decreased environment-focus)" (Northoff, 2014, p. 405; 2016; Feinberg, 2011). Further, the increased self- and body-focus mean that the depressed person's attention is no longer focused on their relation to the social environment and environmental events, as in healthy people, but, rather, on internal functions and their own dysfunction. Increased self-focus seems to correlate with decreased environment-focus" (2016, pp. 83–84).

What Josef Parnas calls "presence" is altered in schizophrenia (Parnas, 2003). "The experience of the world and its objects is no longer accompanied by a prereflective self-awareness. The self that experiences the world is no longer included in that very experience; that is, the self no longer experiences itself as being the self of its own experiences" (Northoff, 2015, p. 85). "The person's own self and the environment are experienced in an altered way with abnormal degrees of self-specificity" (Northoff, 2014, p. 391). Imaging has shown resting state abnormalities in people with schizophrenia. It has shown abnormal functional connectivity and low-frequency fluctuations in midline regions of the brain (391). Because of how the disease alters the processing of information, people with schizophrenia may feel detached or isolated from the world. Their distorted perception of the world may occur in positive symptoms of hallucinations and delusions, as well as negative symptoms of flat affect, anhedonia, and avolition. It impairs their ability to respond to reasons for actions that will enable them to adapt to and engage with the world (Glannon, 2019, pp. 67–83).

In patients with major depression and schizophrenia, disrupted neural and mental processing impairs their capacity for flexible behaviour in meeting the cognitive and physical demands of the environment. Instead, their thought and behaviour become maladaptive in failing to meet these demands. These disorders can impair their ability to be effective agents. Both disorders involve dysfunction in neural networks that mediate psychomotor, cognitive, affective, and volitional interactive resonance necessary to form and execute action plans in response to the environment. In healthy brains, mind and brain are two levels of a unified system that interact in a series of re-entrant loops that maintain homeostasis within the organism and enable it to accurately process information from sources external to it. Disrupted processing due to genetic and epigenetic factors influenced by chronic psychosocial stress and other aversive events results in disrupted planning, choosing, and acting. Their thought and behaviour fail to align with actual events in the world or reflect a highly maladaptive response to such events.

Psychopathy is another disorder that impairs agency in a nuanced and interesting way. Disrupted connectivity between the reward circuitry involving the nucleus accumbens of the ventral striatum and the ventro-medial prefrontal cortex has been shown in functional imaging studies of adult male incarcerated offenders. "These data suggest that cortical-striatal circuit dysregulation drives maladaptive decision-making in psychopathy, suggesting the notion that reward system dysfunction comprises an important neurobiological risk factor" (Hosking et al., 2017, p. 221). When the dysfunction is examined more closely it emerges that an affective defect involving empathy is very deficient although other subjective resonances involving appetites of the self are relatively intact (Gillett, 2009, Ch.9).

Wexler's cultural neuroscientific point about neuro-environmental consonance where the environment is considered ethologically, indicates how social factors such as the trauma of being a political or climate refugee can cause dissonance between one's brain and mind and new environments, are relevant here as well. Internal-external dissonance can disrupt the brain processes mediating mental states and result in depression, anxiety, or other neuropsychiatric disorders involving moderately or severely impaired agency. This dysfunction may occur at any stage of life but may be more likely in earlier and later stages when one is less flexible in responding to changes in one's environment.

These are examples of how neuropathologies can interactively impair brain function and the motor, cognitive, affective, and volitional capacities it mediates. They show that how one adapts or fails to adapt to changing circumstances depends not just on the brain or brain-mind but on interaction between and among the brain, mind, body, and environment. But brain dysfunction alone can disrupt this interaction. Hallucinations and delusions in the positive symptoms of schizophrenia are associated with abnormalities in grey and white matter tracts and disrupted connectivity between the striatum and parietal and frontal cortices. These features can cause distorted perception and interpretation of information from the external milieu and in turn impair executive functions necessary to responds to reasons, make rational decisions, and perform voluntary actions. Aristotle's epistemic condition of ignorance (*agnoia*) of the circumstances of action in Book III of the Nicomachean Ethics, together with compulsion or force (*bia*), undermines or precludes voluntary behaviour (Aristotle, 1984, Volume II, Bk. III). In neuropsychiatric disorders, dysfunction in neural networks mediating beliefs can have this effect. This is especially the case in the positive form of schizophrenia consisting of delusions and hallucinations.

Free will scepticism based on the model of Libet's experiments (Wegner, 2002) assumes that conscious intentions are necessary for free will and that unconscious neural processes preclude it. But unconscious sensorimotor capacities are just as critical as conscious cognitive, affective, and volitional capacities in agency. This is evident in activities such as driving a car or riding a bicycle. The conscious learning through which we acquire these skills becomes unconscious procedural capacity that allows us to perform them. Conscious intention is not the only critical component in free will, or free agency (Roskies, 2010; Searle, 2010). Neuropathologies that impair motor functions and the physical ability to perform voluntary bodily movements are the most obvious examples of conditions that impair intentional and voluntary agency. This is evident in Parkinson's disease. Again, PD involves not only motor impairment but cognitive, affective, and volitional impairment as well. Impairment in any or all four of these components of sensorimotor integration can impair agency to varying degrees. Indeed, the degeneration of neurons in the motor circuits of the basal ganglia (striatum, globus pallidus, and putamen) may also include degeneration of neurons on associative and limbic circuits. Because these circuits project to and receive inputs from the prefrontal cortex, motor impairment may coincide with cognitive impairment in brain regions mediating executive functions.

PD is a good example of how a neuropathology can undermine the ability to plan, initiate, and complete actions. In this disorder, bradykinesia, or slowness of movement, "encompasses more than basic motor acts. The patient often exhibits problems with planning, initiating, and executing coordinated and sequential actions" (McNamara, 2011, p. 17). When PD patients experience bradykinesia from the disorder or other dyskinesias as side effects from dopamine agonists to treat the disorder, the agentic self does not have enough strength to initiate and control actions. Motor programs are intact and available, but the patient has difficulty activating them and controlling them once activated. Activation and control ... "requires an intact agentic self" (McNamara, 2011, p. 17). Varying degrees of motor, cognitive, affective, and volitional impairment can cause different types of paralysis and limit a PD patient's agency.

Lesions in the cerebellum can also impair motor and cognitive control by impairing coordination, balance, and assessment of temporal intervals in motor planning and the timing of actions (Linden, 2019). Among other functions, they can affect proprioception and sensorimotor integration that are necessary for controlled movement in navigating the physical environment. Varying degrees of motor, cognitive, affective, and volitional impairment can cause different types and degrees of physical and mental paralysis and limit agency in patients with Parkinson's and other neuropsychiatric disorders.

In Book II of the *Republic*, in the myth of Gyges Ring, Plato asks why a person who could rape, kill, and commit other heinous acts with impunity might refrain from doing them. By the end of the dialogue, the answer emerges that an individual who committed these acts, driven by certain appetites rather than reason, would have a disordered soul. The neuropathologies that we have described cause a disordered soul in a different sense. They disrupt normal function of the brain-mind-body relation necessary for agency.

Consciousness is an essential aspect of voluntary agency and its neural basis hotly debated. We consciously deliberate about which actions to perform or avoid, consciously form intentions to act, and consciously execute intentions in deciding to act. Yet some neuropathologies show that being too conscious of one's actions can interfere with executing basic motor skills. This phenomenon is popularly referred to as 'over-thinking', but the pathologies can be debilitating. For example, in obsessive-compulsive disorders, excessive rumination or lack of self-confidence in executing motor tasks can prevent one from performing movements voluntarily (De Haan et al., 2015; Kiverstein et al., 2019). A person with ideomotor apraxia can perform motor tasks such as answering a phone call without thinking about doing it but cannot perform them when instructed to do so. In this disorder, the conscious cognitive capacity to respond to commands is impaired by semantic amnesia from brain injury. These are cognitive rather than motor disorders. Although OCD is associated with dysregulated frontal-striatal pathways, and ideomotor apraxia is associated with dysfunctional motor regions of the brain, these disorders illustrate the importance of complementary and interwoven conscious and unconscious processes in effective agency.

Neuropathologies including, but not limited to, those we have described can impair agency to varying degrees by impairing motor mental capacities to varying degrees. They can impair or undermine the capacities of the agentic self. They can also inform normative judgments about moral and criminal responsibility for actions and omissions. The degree of impairment can influence whether individuals deserve full or mitigated moral and criminal responsibility for, or excuse from, their actions. Neuroimaging alone will not determine whether a person's motor, cognitive, affective, and volitional capacities were intact at the time they performed an action because the brain activity that mediates these capacities may change and be very different at the time of a scan than they were at the time of the action. Blood flow, glucose metabolism, and neural oscillation and synchronization are not static but dynamic. Imaging measuring these brain features at one time may be very different from imaging measuring them at another time (Roskies, 2013). Extensive structural and functional damage from brain injury, neurodevelopmental, or neurodegenerative conditions displayed by imaging may correlate with these impairments and dispose judges and juries to excuse agents for criminal acts. These findings may be more likely to indicate that the agent did not have enough control of their action or actions to be fully responsible for them. But determining the degree of impairment may be difficult.

Even in cases of significant neural damage and dysfunction confirmed by a CT or MRI scan, this may not provide conclusive evidence that the person lacked control of their behaviour. Depending on the degree of plasticity in the person's brain, some regions not directly involved in executive functions may take over for other regions that have become dysfunctional. In some cases, the parietal cortex may take over from a damaged prefrontal cortex to enable some of these functions. Similarly, other limbic regions may take over the functions of a damaged anterior cingulate cortex and enable some degree of planning and decision making. A hyperactive amygdala and corresponding impaired executive functions in the PFC (specifically the vmPFC) may be associated with difficulty in controlling one's impulses and thus impair the cognitive and volitional components of behaviour control. This can disrupt normal connectivity between limbic and prefrontal areas and a normal balance between inhibitory and excitatory mechanisms. Disruption associated with a hyperactive amygdala or other limbic areas may cause chronic activation of the fear system and result in anxiety or depression. Imbalance between excitatory and inhibitory mechanisms mediated by hypothalamic-hippocampal circuitry that ordinarily regulates impulsivity can also result in impulsive behaviour or aggressivity and impair affectively modulated agency (Noble et al., 2019).

Other neuropathologies can impair effective agency in other ways. Disrupted connectivity between the amygdala and ventromedial and orbitofrontal prefrontal cortex (OFC/PFC) can impair the anterior cingulate cortex (ACC) -mediated capacity to choose between alternative courses of action and resolve cognitive conflicts. But unless the hyperactivity was severe, imaging would not determine whether a person was unable to control their impulses in acting, or whether they had difficulty controlling them but did not put enough effort into trying to control them. The latter case could be a mitigating condition, but not an excusing one. A hyperactive amygdala might not completely disable inhibitory (deliberative, reflective) mechanisms in the prefrontal cortex to which it projects as one part of cognitive-emotional processing. Some PFC-mediated executive functions may be intact and preserve the mental capacities necessary to control which actions one performs or refrains from performing. Agentic control is not always completely present or absent but may (or even is likely to) come in degrees. These variations can complicate judgments about how much control a person has over their actions and whether they deserve attributions of responsibility, mitigation, or excuse. In most cases, imaging will supplement and confirm behavioural evidence but not supplant it. Imaging alone will not determine whether an individual could or could not control their behaviour at specific times or over time or in relation to particular holistic situations touching personal sensibilities (Morse, 2010; Glannon, 2011, pp. 72ff.).

Neuroimaging may be able to adjudicate conflicting claims about behavioural control when behavioural evidence is ambiguous or inconclusive. In cases where behavioural evidence of impaired reasoning and decision-making is ambiguous, neuroimaging can clarify how the brain mediates the mental states associated with actions. Imaging is limited in this regard. However, because the brain changes over time. Imaging showing certain structural or functional properties of the brain at a later time may be very different from the structure and especially function of the brain at the earlier time when the person acted. There is also a problem with methodology in relation to a holistic dynamic mode of brain function. The standard procedure for functional neuroimaging is contrastive, which is superb for identifying structures differentially involved in a defined function and regarding as background areas active in many functions to some degree. Differentiation of anatomical site by function has been a mainstay of neurology since its beginnings and based on the differential effects of clinical lesions on functional abilities. This has been effective in assembling a caste list of crucial enabling conditions for intact function.

Neurologists are skilled at identifying subtle indicators revealing 'where is the lesion'. This is the stuff of laying the groundwork of medical teaching in clinical neuroscience. It also has a part in functional understanding of integrated neuro-cognitive abilities. Yet, as Luria's patient Zazetsky showed in his will to recover his cognitive functions after his brain injury, restoration of these functions over time reveals a holistic and dynamic top-down trajectory in the understanding of integrated brain function (Gillett & Butler, 2021). Zazetsky's case illustrates that persistence, resilience, motivation, and effort can tell us more about a person's capacity for agency than the effects of natural laws and past events that have figured so prominently in debates about free will.

This is just one feature of the dynamic brain and why imaging showing brain functions in real-time is limited in explaining human behaviour. More revealing about control are diachronic accounts of human agency that include but are not restricted to actions at specific times. These accounts may better approximate an understanding of whether individuals could control their behaviour when they performed a harmful action and whether they could be morally or criminally responsible for what they do or fail to do in specific circumstances.

Even in cases where loss of behaviour control is associated with a neuropsychiatric disorder, an affected person may still be responsible for actions performed in a disordered state if they fail to take steps to prevent being in such a state and know that this failure may result in loss of control of one's actions. The cognitive control they have in knowing the probable outcome at a later time of what they do or intend to do at an earlier time could make them responsible for the outcome if it occurs. For example, a person with Bipolar Disorder I may fail to take a mood stabilizing drug that would prevent mood cycling and mania or hypomania. They may be responsible for actions committed in a manic or hypomanic state because of their negligence in failing to take measures that they know would prevent it. In an actual case, a patient with anxiety and depression received deep brain stimulation to the nucleus accumbens, which ameliorated his symptoms. He asked his psychiatrist to increase the frequency of the stimulation so that he could feel even better. This resulted in euphoria and his comment that the higher level of stimulation made him feel 'exceptionally well'. Yet he retained insight into his condition and expressed concern that the euphoria from the increased frequency might cause him to lose control of his thought and behaviour. He asked his psychiatrist to decrease the stimulation frequency to a lower therapeutic level (Synofzik et al., 2012).

If the patient could increase or decrease the frequency on his own, if he increased it knowing that it could result in hypomanic or manic state, and if he committed a harmful act in this state, then he could be responsible for the action despite being incapacitated when he committed it. He could be responsible on grounds of negligence. In the hypothetical and actual cases that we have described, a person may be responsible for actions committed in a neurally and mentally disordered state if they had cognitive control in knowing the likely outcome of an action, or omission, yet failed to take measures would have prevented dysfunction in this state. Importantly, the narrative considerations we have invoked are the kinds of things that ordinary folk (like a jury) are often sensible about (pun intended).

Neuropathologies as such are not normatively decisive though significant. They have normative implications when they impair the motor and mental capacities that enable behavioural control and agency and in terms of which we praise and blame people and hold them morally and criminally responsible for their actions and omissions. Neuropsychiatric disorders can affect the relevant capacities to varying degrees. This can determine whether they are mitigating or excusing conditions, or whether individuals with these disorders themselves maintain enough control of their behaviour to be fully responsible for what they do or fail to do.

The bases of this complex development, and the neuro-cognitive capabilities which one forms in response to it, are illuminated from an indigenous and post-colonial perspective. We discuss this in Chaps. 4 and 6. This involves a much closer relationship between the subject and the natural and social environment, and a richer sense of adaptability than philosophical perspectives grounded in Anglo-American and Continental perspectives both of which aim at a natural fit for a neuro-philosophy grounded in the rhythms of the brain. Continental philosophy is especially helpful in assessing adaptability and control because it examines not only actions and the specifiable neural processes from which they issue but also the holistic context in which one acts. It explains actions based on the agent's relation to the world in which they live, think, and act. Events both inside and outside the brain influence neural rhythms and how they enable agency and adaptive behaviour. That conceptual exploration is genuinely transformative for philosophy, in that previously the combination of viewpoints has been incorporated into neither philosophy of mind nor post-enlightenment neuro philosophy in general. In fact, post-enlightenment neuro-philosophy has tended to construct itself on the basis of Anglo-American philosophy and to neglect broader insights derived from Continental, post-modern, and indigenous thinkers. This has sadly alienated not only many philosophers but also indigenous communities of scholars and deprived us all of the significant insights that they can bring to a nuanced naturalistic philosophy of mind.

ACTION AND THE NEURAL NET

Descartes' distinction between *res cogitans* and *res extensa* conceals the fact that our ideas and theories about things emerge from our interaction with the world. The idea of 'reality' captures the fact that some of those interactions do not merely reflect how we think counterfactually but are part of our actual ecological/ethological situation.

The neural net is holistic and associative. ART recognises that cause-effect links, construed in the normal way, are nested in a complex and interwoven network of neural associative connections (in which bottom up, diffuse nudges or influences and whole-part or top-down effects all operate). These connections subserve the formation of lifelong patterns of connection modifying inherited tendencies so that individual organismic history and situated experience (not general laws) make an important contribution to an individual's behavioural repertoire. Thus, ART is the *fons et origo* of neurocognitive singularity.

Action is always in a context and an expression of the individual so that in between the experience and the act is an intention—a cognitive and formative creation. This 'fore-shadow' cognitively corresponds to what one is trying to do (which can be 'shady' in a sense different from that evoked by TS. Eliot—or perhaps not). The action description locates an event or series of events in an individual or personal history which itself casts 'shadows' in a creature's life locating that individual in a socio-historical site.

CONCLUSION

A neurodynamic view of the soul as a complex of sensorimotor, cognitive, affective, volitional, and social aspects of human agency and life enables us to take a fresh approach to neuroimaging-based work on cognition in the last decade. While this work shows how the brain mediates the mental states that lead to action, it is limited because images of brain structure and function cannot show how the environment influences the brain and mind. Specifically, a neurodynamic view of the soul as a psychophysical process emphasises that how the brain mediates mental states such as desires, beliefs, intentions, and decisions is shaped by the organism's and subject's natural and social environment. Explaining this process involves not just a brain-mind relation but more complex relations between brain, body, mind, and world. In these respects, we follow Aristotle in focusing on an organic conception of the human psyche. This enables a philosophical exploration of psychology and ethics which breaks new ground by moving beyond the divided dualistic and mechanistic factions in neuro-philosophy.

These considerations indicate that effective (flexible, adaptive) agency, involves not only the mind, or the mind-brain, but mind, brain, body, and environment. But the brain dysfunction in neuropsychiatric disorder can impair agency by impairing motor, cognitive, affective, and volitional capacities necessary to plan, initiate, and complete action plans. This impairment can influence judgments of full responsibility, mitigation, or excuse. More generally, brain dysfunction that interferes with motor and mental capacities can present challenges for practical reasoning and the excellence in achieving individual and collective goals that allows us to flourish in life. As we explain in the next five chapters, and especially in Chap. 7, flourishing is the ultimate manifestation of the normative dimension of human behaviour in knowing how to go on and being good.

A more complete understanding of adaptability requires more discussion of the relation between the individual agents and the physical and social milieu in which they live and act. Continental philosophers of the twentieth century have discussed this relation in terms of the general concept of being-in-the-world. This is the topic of Chap. 3.

REFERENCES

Aristotle. (1984). *Nicomachean ethics.* In *The complete works of Aristotle,* Volume 2. (J. Barnes, Trans. and Ed.). Princeton University Press.

Baumeister, R., Mele, A., & Vohs, K. (Eds.). (2010). *Free will and consciousness: How might they work?* Oxford University Press.

Bechtel, W. (2008). *Mental mechanisms: Philosophical perspectives on cognitive neuroscience.* Taylor & Francis.

Beurel, E., Toups, M, & Nemeroff, C. (2020). The bidirectional relationship of depression and inflammation: Double trouble. *Neuron, 107,* 234–256.

Bratman, M. (1987). *Intention, plans, and practical reason.* Harvard University Press.

Bratman, M. (2007). *Structures of agency: Essays.* Oxford University Press.

Bratman, M. (2014). *Shared agency: A planning theory of acting together.* Oxford University Press.

Bressler, S., Kay, L., Kozma, R., Liljenstrom, H., & Vitiello, G. (2018). Freeman neurodynamics: The past 25 years. *Journal of Consciousness Studies, 25,* 15–32.

Buzsaki, G. (2006). *Rhythms of the brain.* Oxford University Press.

Buzsaki, G. (2019). *The brain from inside out.* Oxford University Press.

Buzsaki, G., & Draguhn, A. (2004). Neural oscillations in cortical networks. *Science, 304,* 1926–1929.

Buzsaki, G., & Freeman, W. (2015). Editorial overview: Brain rhythms and dynamic coordination. *Current Opinion in Neurobiology, 31*(2), v–ix.

Castle, D., & Buckley, P. (2015). *Schizophrenia* (2nd ed.). Oxford University Press.

Christensen, A. L., Goldberg, E., & Bougakov, D. (Eds.). (2009). *Luria's legacy in the 21st century.* Oxford University Press.

Clarke, R. (2003). *Libertarian accounts of free will.* Oxford University Press.

Craver, C. (2007). *Explaining the brain: Mechanisms and the mosaic Unity of neuroscience.* Oxford University Press.

Davidson, D. (2001). *Essays on actions and events* (2nd ed.). Clarendon Press.

De Haan, S., Rietveld, E., & Denys, D. (2015). Being free by losing control: What obsessive-compulsive disorder can tell us about free will. In W. Glannon (Ed.), *Free will and the brain: Neuroscientific, philosophical, and legal perspectives* (pp. 83–102). Cambridge University Press.

Dennett, D. (2003). *Freedom evolves.* Penguin Books.

El Haj, M., & Lenoble, Q. (2018). Eying the future: Eye movement in past and future thinking. *Cortex, 105,* 97–103.

Evans, G. (1982). In J. McDowell (Ed.), *The varieties of reference.* Clarendon Press.

Feinberg, T. (2011). Neuropathologies of the self: Clinical and anatomical features. *Consciousness and Cognition, 20,* 75–81.

Fischer, J. M., & Ravizza, M. (1998). *Responsibility and control: A theory of moral responsibility.* Cambridge University Press.

Fordham, B., Sugavanam, T., Edwards, K. et al. (2021). The evidence for cognitive behavioural therapy in any condition, population or contecxt: A meta-review of systematic reviews and panoramic meta-analysis. *Psychological Medicine, 59*, 21–29.

Freeman, W. (2000). *Neurodynamics: An exploration in mesoscopic brain dynamics*. Springer.

Freeman, W. (2001). *How brains make up their minds*. Columbia University Press.

Freeman, W., & Changeux, J. P. (2015). *Dialogue on intentionality with Jean-Pierre Changeux*. Video recorded at the conference "Towards a Science of Consciousness." Helsinki, June 2015. http://www.sigtunastiftensen.se/humaaate/

Freeman, W., & Vitiello, G. (2016). Matter and mind are entangled in two streams of images guiding behaviour and informing subjects through awareness. *Mind & Matter, 14*, 7–24.

Fuchs, T. (2018). *Ecology of the brain: The phenomenology and biology of the embodied mind*. Oxford University Press.

Fuchs, T. (2021). *In defense of the human being: Foundational questions of embodied anthropology.*Oxford University Press.

Fuster, J. (2013). *The neuroscience of freedom and creativity: Our predictive brain*. Cambridge University Press.

Gallagher, S. (2005). *How the body shapes the mind*. Oxford University Press.

Gallagher, S. (2020). *Action and interaction*. Oxford University Press.

Gennaro, R. (ed.) (2015). *Disturbed consciousness: New essays on psychopathology and theories of consciousness*. MIT Press.

Gillett, G. (2009). *The mind and its discontents* (2nd ed.). Oxford University Press.

Gillett, G. (2015). Evolution, dissolution and the neuroscience of the will. In W. Glannon (Ed.), *Free will and the brain: Neuroscientific, philosophical, and legal perspectives* (pp. 44–65). Cambridge University Press.

Gillett, G., & Butler, M. (2021). When the music's over, then dancing with a partner will help you find the beat. *Cambridge Quarterly of Healthcare Ethics, 30*, 631–636.

Gillett, G., & Webster, D. (1975). Mediodorsal nucleus and behavior regulation in the rat. *Physiology and Behavior, 14*, 883–885.

Glannon, W. (2011). *Brain, body, and mind: Neuroethics with a Hunan face*. Oxford University Press.

Glannon, W. (2019). *Psychiatric Neuroethics: Studies in research and practice*. Oxford University Press.

Goldapple, K., Segal, Z., Garson, C. et al. (2004). Modulation of cortico-limbic pathways in major depression. Treatment-specific effects of cognitive behavior therapy. *Archives of General Psychiatry, 61*, 31–41.

Grossberg, S. (2013). Adaptive resonance theory: How a brain learns to consciously attend, learn, and recognize a changing world. *Neural Networks, 37*, 1–47.

Haggard, P. (2008). Human volition: Towards a neuroscience of will. *Nature Reviews Neuroscience, 9*, 934–946.

Haggard, P. (2011). Does brain science change our view of free will? In R. Swinburne, ed., *Free will and modern science*. Oxford University Press, 7–24.

Hallett, M. (2007). Volitional control of movement: The physiology of free will. *Clinical Neurophysiology, 11*, 79–92.

Hallett, M. (2016). Physiology of free will. *Annals of Neurology, 80*, 5–12.

Haynes, J.-D. (2010). Beyond Libet: Long-term prediction of free choice s from neuroimaging signals. In W. Sinnott-Armstrong & L. Nadel, eds., *Conscious will and responsibility: A tribute to Benjamin Libet*. Oxford University Press, 85–96.

Heckers, S., & Walther, S. (2023). Catatonia. *New England Journal of Medicine, 389*, 1797–1802.

Hohwy, J. (2013). *The predictive mind*. Oxford University Press.

Hosking, J., Kastman, E., Dorfman, H., Samanez-Larkin, G., Baskin-Sommer, A., Kiehl, K., et al. (2017). Disrupted prefrontal dysregulation of striatal subjective value signals in psychopathy. *Neuron, 95*, 221–231.

Kane, R. (1996). *The significance of free will*. Oxford University Press.

Kircher, T., & David, A. (Eds.). (2003). *The self in neuroscience and psychiatry*. Cambridge University Press.

Kirmayer, L., Lemelson, R., & Cummings, C. (Eds.). (2015). *Re-visioning psychiatry: Cultural phenomenology, critical neuroscience, and global mental health*. Cambridge University Press.

Kiverstein, J., Rietveld, E., Slagter, D., & Denys, D. (2019). Obsessive-compulsive disorder: A pathology of self-confidence? *Trends in Cognitive Sciences, 23*, 369–372.

Laeng, B., Nabit, S., & Kitaoka, A. (2022). The eye pupil adjusts to illusorily expanding holes. *Frontiers in Human Neuroscience, 16*, 877249. https://doi.org/10.3389/fnhuman.2022.877249

Libet, B. (1985). Unconscious cerebral initiative and the role of conscious will in voluntary action. *Behavioral and Brain Sciences, 8*, 529–566.

Libet, B. (2004). *Mind time: The temporal factor in consciousness*. Harvard University Press.

Liljenstrom, H. (2018). Intentionality as a driving force. *Journal of Consciousness Studies, 25*, 206–229.

Liljenstrom, H. (2022). Consciousness, decision making and volition: Freedom beyond chance and necessity. *Theory in Biosciences, 141*, 125–140.

Linden, D. (2019). *The biology of psychological disorders* (2nd ed.). Springer/Red Globe Press.

Loux, M. (Ed.). (1980). *The possible and the actual: Readings in the metaphysics of modality.* Cornell University Press.

Luria, A. R. (1973). *The working brain: An introduction to neuropsychology* (B. Haigh, Trans.). Harvard University Press.

Luria, A. R. (1987). *The man with a shattered world: The history of a brain wound* (L. Solotaroff, Trans.). Harvard University Press.

McDowell, J. (1998). *Meaning, knowledge, and reality.* Harvard University Press.

McNamara, P. (2011). *The cognitive neuropsychiatry of Parkinson's disease.* MIT Press.

Mele, A. (2009). *Effective intentions: The power of conscious will.* Oxford University Press.

Mele, A. (Ed.). (2020). *Surrounding self-control.* Oxford University Press.

Miles, K. (2015). Can we create an artificial brain? *Huffington Post,* September 16. https://www.huffpost.com/entry/artificial-brain-n-55171ebfe4b042295e372e90.

Morse, S. (2010). Lost in translation? An essay on law and neuroscience. In M. Freeman (Ed.), *Law and neuroscience: Current legal issues* (pp. 529–562). Oxford University Press.

Morse, S., & Roskies, A. (Eds.). (2013). *A primer on criminal law and neuroscience.* Oxford University Press.

Nasser, H., Calu, D., Schoenbaum, G. et al. (2017). The dopamine prediction error: Contributing to associative models of reward learning. *Frontiers in Psychology, 8,* 244. https://doi.org/10.3389/fpsyg.2017.00244.

Noble, E., Wang, Z., Liu, C., et al. (2019). Hypothalamus-hippocampus circuitry regulates impulsivity via melanin-concentrating hormone. *Nature Communications, 10,* 4923.

Northoff, G. (2014). *Unlocking the brain: Volume 2—Consciousness.* Oxford University Press.

Northoff, G. (2015). How the self is altered in psychiatric disorders: A neurophenomenal approach. In L. Kirmayer, R. Lemelson, & C. Cummings (Eds.), *Re-visioning psychiatry: Cultural phenomenology, critical neuroscience, and global mental health* (pp. 81–116). Cambridge University Press.

Northoff, G. (2016). *Neuro-philosophy and the healthy mind: Learning from the unwell brain.* W.W. Norton & Company.

Northoff, G. (2018). *The spontaneous brain: From the mind-body to the world-brain problem.* MIT Press.

Northoff, G. (2021). Why is there sentience? A temporo-spatial approach to consciousness. *Journal of Consciousness Studies, 28,* 67–82.

Parnas, J. (2003). Self and schizophrenia: A phenomenological perspective. In T. Kircher & A. David (Eds.), *The self in neuroscience and psychiatry* (pp. 217–241). Cambridge University Press.

Powell, A., Gallur, L., Koopowitz, L., & Hayes, M. (2020). Parkinsonism in the psychiatric setting: An update on clinical differentiation and management. *BMJ Neurology Open, 2*, e000034. https://doi.org/10.1136/bmjo-2019-000034

Roskies, A. (2010). Freedom, neural mechanism, and consciousness. In R. Baumeister, A. Mele, & K. Vohs (Eds.), *Free will and consciousness: How might they work?* (pp. 153–171). Oxford University Press.

Roskies, A. (2013). Brain imaging techniques. In S. Morse & A. Roskies (Eds.), *A primer on criminal law and neuroscience* (pp. 37–74). Oxford University Press.

Rowlands, M. (2010). *The new science of the mind: From extended mind to embodied phenomenology.* MIT Press.

Schacter, D., & Addis. D.R. (2007). The cognitive neurocience of constructive memory: Remembering the past and imagining the future. *Philosophical Transactions of the Royal Soceity B: Biological Sciences, 362,* 773–786.

Schacter, D. Addis, D.R., Hassabis, D. et al. (2012). The future of memory: Remembering, imagining and the brain. *Neuron, 76,* 677–694.

Schultz, W. (2017). Reward prediction error. *Current Biology, 27,* R369–R371.

Schultz, W., Dayan, P., & Montague, P. R. (1997). A neural substrate of prediction and reward. *Science, 275,* 1593–1599.

Searle, J. (2010). Consciousness and the problem of free will. In R. Baumeister, A. Mele, & K. Vohs (Eds.), *Free will and consciousness: How might they work?* (pp. 123–134). Oxford University Press.

Sinclair, A., Manalilli, G., Brunec, I. et al. (2021). Prediction errors disrupt hippocampal representations and update episodic memories. *Proceedings of the National Academy of Sciences, 118,* e2117625118. https://doi.org/10.1073/pnas.2117625118.

Spector, S., Tranah, A., Cull, C., & Goldstein, L. (1999). Reduction in seizure frequency following short-term group intervention for adults with epilepsy. *Seizure, 8,* 297–303.

Spence, S. (2009). *The actor's brain: The cognitive neuroscience of free will.* Oxford University Press.

Strakowksi, S. (2014). *Bipolar disorder.* Oxford University Press.

Strawson, G. (2010). *Freedom and belief* (revised edition). Clarendon Press.

Synofzik, M., Schlaepfer, T., & Fins, J. (2012). How happy is too happy? Euphoria, neuroethics, and deep brain stimulation of the nucleus accumbens. *AJOB-Neuroscience, 3*(1), 30–36.

Velleman, D. (1992). What happens when someone acts? *Mind, 101,* 461–481.

Wegner, D. (2002). *The illusion of conscious will.* MIT Press.

Wexler, B. (2008). *Brain and culture: Neurobiology, ideology, and social change.* MIT Press.

Woodward, J. (2021). *Causation with a human face: Normative theory and descriptive psychology.* Oxford University Press.

Wong, A., Devason, A., Umana, I. et al. (2023). Serotonin reduction in post-acute sequelae of viral infection. Cell 186, https://doi.org/10.1016/j.cell.2023.09.013.

Zeman, A. (2004). *Consciousness: A user's guide.* Yale University Press.

Zeman, A. (2008). Does consciousness spring from the brain? Dilemmas of awareness in practice and in theory. In L. Weiskrantz & M. Davies (Eds.), *Frontiers of consciousness.* Oxford University Press.

Zeman, A. (2014). Neurology is psychiatry—and vice versa. *Practical Neurology, 14*, 136–144.

Zhang, Y., Roy, D., Zhu, Y., et al. (2022). Targeting thalamic circuits rescues motor and mood deficits in PD mice. *Nature.* https://doi.org/10.1038/s41586-022-04806

CHAPTER 3

Being in the World (After Wittgenstein)

Wittgenstein began his philosophical analysis of thought and the mind (or *psyche*) with the philosophical logic of Gottlob Frege (1892–1918/1980) and Bertrand Russell (1905, 1912, pp. 23ff, 66ff.). They had worked on the analysis of language according to its logico-mathematical structure and by that route arrived at a formal conception of human thought (Frege, 1892–1918/1980). This analysis resulted in the *Tractatus Logico-Philosophicus* (Wittgenstein, 1921/1974) which Wittgenstein himself later transcended by introducing thoughts consistent with anthropology and the emerging scholarship in psychology. Alerted by his own earlier attempt to analyse natural language in a 'scientific' and highly structured way, he developed a healthy scepticism about language and meaning which distanced him from his philosophical sponsor, Russell. The early Wittgenstein was thus at the forefront of a new and seemingly rigorous approach to philosophy which assimilated it to the scientific image and his mentors' ideals. The later Wittgenstein was more inclusively naturalistic.

Like Wittgenstein, some current philosophers have been forced to move beyond a narrow reductive materialism. This is similar to the reductive thinking that grounds industry, engineering, and enlightenment scientific naturalism. They have done this through their interaction with [postmodern and continental thinkers, ethologists such as Konrad Lorenz (1952, 1966), anthropologists such as Peter Wilson (1980, 1988), and social theorists such as John Shotter (1984). Thereby philosophers have

G. Gillett, W. Glannon, *The Neurodynamic Soul*, New Directions in Philosophy and Cognitive Science, https://doi.org/10.1007/978-3-031-44951-2_3

arrived at a more nuanced naturalism to debate. The human orientation of this naturalism and the sophisticated ethology of early modern and continental thought led not only to existentialism but also the anthropologically informed conception of 'rule following' in Wittgenstein's *Philosophical Investigations, On Certainty*, and *Culture and Value* (1998). In addition, the ground-breaking neuropsychology of Luria (1973) and Lev Vygotsky (1978), the cognitive psychology of Ulric Neisser (1967, 1976), and the philosophical and social psychology of Rom Harre (1984) have transformed psychology from a narrow form of enlightenment science to a discipline which can embrace nuanced social theory, on the one hand and complexity theory, on the other. That transformation has forced a rethink in the philosophy of psychology and neurophilosophy.

The neuroscience that followed Luria and Vygotsky also offered a dynamic develop-mentalism that replaced the abstract cognitive doctrines that had been modelled on linear causal explanation and reductively scientific or mathematically framed thought. These earlier doctrines so influenced Jean Piaget and his followers regarding the stages of cognitive development and thinking about how we construct mental models, or 'schemata' of the world, that the discursive element of child development was de-emphasised (Piaget, 1929, 1974, 1999). The delay of a corrective move in academic thought was socio-politically understandable, as Vygotsky and Luria were Russian, and the huge growth in scientific and psychological thought was centred in the USA and UK. The transfer of the new ideas and their uptake was discouraged and also hampered by those socio-political forces (as indeed continental thinkers such as Nietzsche would have predicted).

In the neglected framework, language and cognition incorporated a socially informed understanding of human life that revealed various styles of thinking. These styles, as noted, added an anthropological and socio-political authenticity to the resulting neurocognitive theory. But subsequent neurophilosophy has sometimes lost touch with those expanded naturalistic insights. This is despite the fact that such insights have been increasingly influential in educational practice and based on a more widely grounded variety of psychological theory incorporating discourse and its subtleties. One could speculate whether this neglect might be influenced by the economic power and financial interests of the 'scientific-industrial' sector and their role in modern society (and modern medicine). These powerful interests may go some way to explaining the post-industrial formalisation of philosophy grounded in a mathematical and scientific model rather than actual human behaviour as it exists in the real and complex world.

THE SCIENTIFIC AND SEMANTIC SHAPE OF COGNITION AND THOUGHT

Natural science overwhelmingly structures itself in terms of linear causal relations. It grounds philosophy of thought by eliminating any dualistic distractions (seen as a residue of religious influence) in trying to establish a two-place relationship between 'statements' (or causally specifiable brain states 'corresponding to' them) and the world. The analytic simplicity depends upon an apparently straightforward philosophical conception of the causal relation between the subject and a complex condition or state of affairs in the environment (Woodward, 2021, Ch. 2) This condition can be all too easily abstracted as the relationship of 'reference' and be rendered in completely extensional (rather than existential) terms. This is a standard form of austere scientific naturalism. Whereas this might be appropriate for causally effective extensional objects construed according to post-industrial abstract metaphors and their associated codes (Fodor, 1987), it hardly seems appropriate for a complex cultural product such as statements in a natural language used and formed by a subject cognisant of cultural, aesthetic, or socio-political realities.

Let us begin with a simple intuition that appears uncomplicated at first sight. "All statements are true or false on the basis of their correspondence or otherwise with the world". This is the conceptual and logical basis of "the slingshot argument" by which some theorists in the philosophy of language and thought have been tempted in developing models of meaning and understanding (Barwise & Perry, 1984, pp. 25ff.; Neale, 2001; Davidson, 1969). This is a particularly stark form of correspondence theory. It marginalises any conceptual equivalent of 'sense' in the interest of some kind of 'direct realism' that renders in stark terms the relationship between sentences thought of as physical sound patterns or some metaphysical cognate expressing truth-claims about the world. One weasel word here is 'correspondence' and another equally tendentious signal of a stark naturalism is the term 'the world'. Both expressions hide a multitude of complexities which Wittgenstein examined in his later philosophy.

The reality giving rise to this rhetorically simplistic and mechanistic abstraction is far more complex and involves the holistically interwoven rhythms of human life as they affect language and inflect psychology. Wittgenstein's becoming alert to this complexity and the actual living natural realities of human life and culture caused him to transition from the philosophically tidy logical and linguistic analysis of the *Tractatus* to the much more nuanced and complex analysis of the *Investigations*. Once a simple two-place relationship has been philosophically sketched in principle, the stage is set for one object to be the state of a cognitive individual and the other to be a state of affairs, both specified and characterised in extensional terms. The appeal is seductive, but misleading.

"Alas, poor Yorick! I knew him, Horatio", says a melancholy Hamlet informed of the identity of the person whose skull has been disinterred. He uses these words after the event when reminded of a character from his childhood and immediately recalls the exercise of 'making sense' of the world and human life (as best a child can) and of a living being who was an important part of that lived autobiography, and who was remembered in the context of warm childhood experiences infused with affective nuances. What we have in the slingshot argument is not even the skeletal remains of a scholarly approach to language but a tendentious 'policy prospectus' or sparse 'promissory note' which gestures at a living and engaging dynamic phenomenon—language as it forms part of the human world.

The two-place theory involving 'language' and 'the world', with the only modification of the first term as a 'referent' reckoned to be causal, mechanistic, or logico-mathematical in its 'functional role', is a quasi-scientific philosophical simplification of a human and complex reality. This causal and extensional basis for thought (as expressed in 'the slingshot argument'), was a neo-Fregean project in reaction against any development of 'sense' as a component of linguistic meaning. It was driven by the idea that 'sense', as a philosophical entity, tended towards a kind of idealism or sensibility which rendered language and thereby logic 'subjective'. But the formalising move abstracted the analysis of language away from a naturalistic investigation of human thought and behaviour into one which could be thought of in terms of quasi-mathematic philosophical logic as an idealistic invention *par excellence*. That abstraction neglects the realities of complex neurocognitive function as the basis of our mode of interacting with the world and others.

It is time to reinstate 'sense' and even 'sensibility' in a way which would, I am sure, meet with approval from both the 'Austin's: male and female (J.L. and Jane) (Austin, 1962). To do such is to rehumanise words and the discourse in which they are embedded. The abstraction is dubious even in terms of human action as distinct from mechanical or causal movements of the human body because actions carry a presumption of intention analysed in 'cognitive' terms, though, in fact, the cognition involved is more complex and continuous with animal adaptation. The stage is set for the rhythms of life to-step into view.

'Intention' entangles action in the holistic, embodied, and situated human associative network of intersecting patterns in the day-by-day reality of human ways of going on (Shotter, 1996; Gallagher, 2020). The quasi-scientific simplification of language may appear to work for denotation. But it does not include varieties of sensibility and interaction that can be descriptively, adverbially, and culturally or anthropologically differentiated from each other. Those many contexts force us to be cognizant of the manner in which we become acquainted with objects (or quasi-objects such as magnetic fields) or even extensional states of affairs with which we are in contact through some complex neuro-cognitive state. Thus, we might say: 'The wind is strong today; it is almost blowing me over'. The sentence is instantly understood by both speaker and hearer, and it is taken to denote the wind as I experience it around here at the present time (and for a person with a stroke, the statement can be taken literally and is known to one of the authors by acquaintance!). Given those perceptual and spatio-temporal restrictions, there is a referent being assessed for one of its qualities and cast into a dynamic use of language connecting speaker and hearer through the ecologically situated and engaged speech acts or social interactions between them.

Even such apparently straightforward terms such as 'all' are deceptive (Wittgenstein, 1983, pp. 42ff.). One can imagine an older experienced woodsman saying to his younger apprentice "Cut all those trees down!" And coming back later to find she had left one standing. When asked "Why did you not do what I said?", she answers "I did, that is not a tree, it is only a sapling". One might be inclined to think 'she has got a point' and admire the way in which her sensibilities have influenced her observation (in both senses of that word).

A simple causal theory of reference for names and other terms omits such a fluid, cultural, and essentially embodied conception of 'sense' as it occurs in nature as an inhabited and humanised context. This was not properly analysed by Frege, Russell, Saul Kripke (1980), Gareth Evans (1982), or indeed any of their quasi-mathematical followers in philosophical logic and philosophy of language. The latter especially inspired a swathe of students at Oxford and in the USA. John McDowell was particularly influential in this regard. However, he was 'midwife' to a richer family who realised that Wittgenstein initially expressed a purified form of this approach in the *Tractatus* but then transcended it in the *Investigations* and all his later philosophy.

In part, the rejection of a broader interpretation of 'sense' was based on a dualistic reading of that term. This could be associated with Locke (following Descartes) or later more extreme idealists who had been relatively sidelined by more scientific thinkers in the British tradition. That was a move in favour of causal naturalists and a post-industrial metaphor for neuro-cognition. The dismissal of any less functional and more dynamic naturalism as being a form of idealism associated with Descartes, Locke, and other earlier modern philosophers pervaded Anglo-American Philosophy for much of its recent history. It is time to redress that neuro-philosophical astigmatism more comprehensively.

The reduced scope of naturalistic thinking ultimately led to various types of reductionism about human thought—both formal and scientific. The first of these reductionisms focuses on the formal relationships between elements in a structured system like language and treats them as arranged into a code where abstract mathematical functions link the elements of the system and ultimately 'the slingshotters' (Fodor, 1987). This scientific reduction dealt with so-called 'states' and 'functions' explained in causal terms. It was a way of dissecting and effectively 'explaining away' or omitting reference to the dynamic processes of life and thought. This approach suggested that thought could be isolated from persons, as if awaiting study in a way something like "a patient etherised on a table" (Eliot, 1915/2002).

THE SOCIAL CONTEXT OF MEANING

Philosophers who have examined the semantics of Fregean 'sense' and predication failed to provide a convincing philosophical account of 'the hurley-burley of life' (Shotter, 1996). This goes beyond attempts to explain thought in terms of linguistic analysis. It requires neuro-philosophical attention to ongoing adaptation and organic reality in a changing ethological environment. The need for fluid and flexible understandings of neurocognitive abilities, liberated from constraining mechanistic metaphors, naturalised and reified abstract logic and the resulting semantics was made evident through dynamic studies of living organisms which Wittgenstein might have found in Aristotle had he sought them there rather than followed Russell and Frege exclusively into Plato and abstraction. However, that move was perhaps fuelled partly by the histori- cal movement beyond Aristotelian or Thomistic thought which had shaped the core of late medieval theistic naturalism.

This latter formal or scientific turn excludes rather than includes men and women in historical contexts whereas the neglected strand means that their neurological function resists simplification as the finished work of a Creator. A more evolutionary view was espoused and explained in the works of the neurologist and psychiatrist John Hughlings Jackson (1887) and the aforementioned neuropsychologist Luria (1973). Interestingly, Hughlings Jackson's work preceded the works of Frege, Russell, and their successors in the philosophy of language. He developed his account of thought and behaviour before the deflection into analytic abstraction by the philosophers of language in the early twentieth century. Luria devel- oped his account before a second deflection later in that century. While neuroscientists and philosophers have explained the neuroscientific anom- alies generating the different 'discontents of the mind', there remains an explanatory gap regarding the nature and content of thought in healthy minds. This gap needs to be filled by neurodynamics and more inclusive neurophilosophy taking account of the complex rhythms of life in the human world (Gillett, 2009; Northoff, 2016). Working separately and at different times in England and the USSR, Hughlings-Jackson and Luria provided the ideas that subsequently would fill this gap (Luria's work became internationally recognised in the 1950s).

They lead us away from mechanistic devices as models of the mind and towards a reconciliation with conceptions of evolved higher animal func- tion and socio-cultural reality. They realised that the mammalian cerebral cortex subsumed adaptive neural circuits under higher and more complex levels of integrative control. The work of Hughlings-Jackson and Luria subsequently influenced the dramatic scientific revolution that was Freeman-style neurodynamics. If it had been properly heeded, then we would not have needed to reconceive neuro-cognitive thought and behav- iour in studies of living and intact brain function and complexity theory, whereby the function of an organism or system cannot be explained in terms of its constituent parts. This re-conception was required to deal with the neural networks of advanced species. The post-industrial tendency to adopt a 'head in the sand' (or 'head in the machine') stance to dynamic neurophilosophy has only recently been rejected in favour of study and analysis articulating and researching a dynamic approach to neurocognition. This recent approach is based on actual living brain function and advanced neuroimaging rather than an approach using computer-based simulations of psychology and intellectually more tractable neuro-anatomical localisations of anthropological function (Han & Northoff, 2008; Northoff & Hayes, 2011).

Northoff offers what he calls a 'category theoretical' approach to consciousness', drawing on extensive dynamic neuroimaging and a complex mathematical analysis (2014, 2018, 2021). The emphasis may appear to many post-industrial analysts to focus on chaotic complexities of neuronal firing in the human neural net as it engages in various tasks. However, what appears as 'chaos' from one perspective might resolve itself into a thinkable 'shape' from another.

There is a complex relationship between his analysis, loosely based on dynamic spatiotemporal self-localisation of neural function, and the much earlier view involving "the generality constraint" (Evans, 1982). Evans argues that our most basic referential language is the demonstrative, and that demonstratives pick out objects to which we have become spatiotemporally proximate. Context and cognitive preparedness interact in a complex associative way. The figure that emerges is cognitively salient according to one's adaptation to a context. This neurocognitive and linguistic activity involves using demonstratives beginning with 'this' or 'that' so as to bring an object or event to 'centre stage' (Gillett, 1992, pp. 166ff). Proper names tend to arise from objects or persons with whom we become acquainted. They may, however, also denote a much more complex referent such as 'The Tudor dynasty,' or 'Gaudy,' or 'Founder's Day,' or the election result.

Understanding predicative terms and phrases as adding information about denoted items in a shared world of activities completes the basic structure on which the complexities of language can be built. This enables us to convey complex information about that shared world to other co-linguistic creatures. It provides a robust and potentially inclusive account of the neurocognitive grounding of thought, especially if one adopts an open and sympathetic attitude to the 'complexity/chaos theory' embedded in contemporary neuro-philosophy (Gillett, 1992, pp. 172ff.).

Anthropology, as in the work of Peter Wilson, confirms this more holistic and nuanced form of naturalism. Wilson authored a series of influential publications including *Man the Promising Primate: The Conditions of Human Evolution* (1980) and *The Domestication of the Human Species* (1988). He explored the conditions and evolutionary developments which have made human beings an apex species in many different ecological environments. As might be expected, these turn out to rest on a number of psychosocial developments which enhance the evolutionary capabilities inherited biologically. These cognitive aspects of human adaptation include human—indeed primate—nature being inherently social, and our facilitating that social function by developing grammatically articulated natural language as a means of communicating successful strategies and of coordinating our efforts. Being able to be detailed in our communications about what is 'out there' lifts us above our natural cohabitants, including the other higher primates.

Complex social conventions with their associated expectations in terms of socio-political and other action in the shared world enhances our adaptation and gives us new ways of going on. Wittgenstein recognized and discussed this in his *Philosophical Grammar* (1991). In this way, the development of such things as neighbourliness, anticipation of challenges, and the coordination of responses was increased. Shared vision (actual and counterfactual), and such shared cognitions as the many imagistic variations on ideas such as the thought that life is eternal and the world full of sources of help and social support, have engendered diverse structures of social power centred on and deriving from a God or Gods.

These ideas loom large in successful societies down through human history. In some societies, such as that of the post-Christian Pacific and Māori or indigenous North Americans, theistic world views incorporating what for Europeans were more or less intellectually and theologically earlier or mythical ideas were both formally and informally embraced. These early ideas introduce historicity into our rhythms of life and initiate a study of human nature as a consciously directed self-forming creature who can benefit from abstract thought and notional aspects of experience such as those in dreams and visions. This inclusive perspective brings an openness to what was originally a hunter-gatherer lifestyle in which small groups developed and were supported by a limited ecological niche or territory in the natural world (Sissons, 2006). On the basis of that trend, urbanisation and the complexities of contemporary life have emerged with all the challenges and innovations they have brought in their wake such as technological ability and its allure.

The language of post-industrial functionality has, however, come to dominate science and academic medicine. The mildly amusing quip "Happiness is a smiling cardiologist" when uttered by a cardiac and stroke patient (such as one of the authors) is immediately meaningful to any thoughtful speaker of English. However, were we to translate that into a designation of the particular biological organisms involved and its complex psycho-neurocognitive or bodily states, it would become completely incomprehensible. One can imagine the sort of statement that would be involved. 'Any Human individual (Hn) is in human neurocognitive state 'Yp' when another human individual with an history of the form Hq and a socio-political specification SPr'. Here a number of specifications of individuated objective times and universally related terrestrial spatio-temporal designations follow the spatio-temporal reference. The difficulty in specifying assessable truth claims multiplies exponentially as we seek to individuate such a situation as the smiling cardiologist in austere extensional terms.

For a related reason, Wittgenstein rejected a metaphysics such as that which informs the slingshot theory and which he had attempted to outline in the *Tractatus*. He discussed the work with Piero Sraffa in Cambridge. These discussions were critical to Wittgenstein's transformation from the "picture theory" of meaning in the *Tractatus* to the "language game" theory of meaning in the *Investigations* (Sinha, 2021). Sraffa is said to have made a common Neapolitan gesture in which the hand is held beneath the chin and the fingers fanned out palm open and nails down and to have quipped 'What is the logical form of that?' This 'form' is supposed to be necessary to explain the adaptations or techniques which guide us through spatio-temporal contexts and the challenges shaping the rhythms of our lives. Clearly it is nothing but an abstract derivative of the reality. It is a small step to relate the resulting post-modern naturalism to Wittgenstein's later notion of "forms of life" as the basis of the resulting dynamic neuro-cognitive function.

Within our historically situated life contexts, language becomes part of our complex biopsychosocial functioning in the world. It is part of a rich neuro-cognitive stream of activity which is captured by our dealings with things and their location as target objects within our lexicon. This lexicon serves as a framework in which items can be discursively located according to our multifaceted ways of going on, of communicating with each other and through which we relate to each other in the world. Frege, as we have noted, conceptualised this discourse and the relationship between its terms as involving both 'sense' and 'reference', where the latter was a physical object or situation denoted by a term or sentence, designating or linguistically capturing a state of affairs. The former term 'sense' seemed to flirt with idealism when associated with the mental accompaniments of language. But this association neglected the multiple ways in which one might neurocognitively relate to a referent. This could be as a line, a marker on a football pitch, a boundary for a zone of non-encroachment, a point to measure, or an indication of a path to follow to a desired destination. Each involves ideas, but none is idealistic because they relate to things we do in the real world. Just as the simple visual indicator formed by a line may have many uses, an acoustic signal (such as a sound or utterance) may in fact have some other use than mere indication in a non-grammatical action of something analogous to visual pointing as an example of referring.

It is easy to discern the origins of the misleading 'slingshot argument' (Davidson, 1969). Stephen Neale linked this argument to attempts by Donald Davidson and others to 'naturalize' the philosophy of language (Neale, 2001). The slingshot argument offers purely "physical or extensional relata: structured sounds or visual stimuli (text) and 'the world'" (Neale, 2001, p. 12). In contrast, reading Wittgenstein enjoins us to share a healthy scepticism about 'scientism'. This scepticism sets one on a path back toward understanding the complexities of our neurocognitive being from a philosophical point of view.

The shift to an understanding of mental and neural activity occurring within a social and cultural context highlights the complexity of language and its constructions. Among other things, it individuates particular ways of thinking of an object, for example, as a stone, a sample of lignite, a piece of Loch Lomond, or a deadly projectile used in homicide (or godly action, as in 'David and Goliath'). The meaning and psychological designations of the object concerned will vary and have differing roles to play in the different human exchanges marked by the language used and context in which the term appears. The varying aspects of events and objects that are picked out by the widely varying words and descriptions we use will be variously relevant to their meaning for us. These words and descriptions can capture the weight, the presence on the surface of human biological substances, the chemistry, the geographical origin, and as many other characteristics as language can convey and for which the imagination can find significance or a role in our forms of life at this point in human history.

For different groups of people engaged in different linguistic activities and seeking to make different cognitive determinations of what they encounter in the world, the words and descriptions they used will have different roles in their subsequent behaviour. This depends on how the object is nestled among their cognitive networks and subsequent activity. Their neurocognitive mechanisms will be as ramifying and complex as their life contexts and the socio-cultural networks in which they operate and form their varied ways of being. This wider perspective allows us to enter the fraught philosophical discussion of consciousness and the intense qualia debate that it has provoked.

We might begin by noticing that there are two uses of ascriptions of consciousness:

1. Consciousness of (x)
2. Consciousness (*simpliciter*)

The former is an entree that allows us to neuro-philosophically approach the latter. When I become conscious of something, for instance a smell in the room, something which has tweaked my sensorimotor interaction with the world appears on centre-stage of my awareness. Thus, the full range of my cognitive resources, including assignments of significance to what I have detected, can be brought into action. For instance, I might conclude that what I have faintly detected is that the bowl in the kitchen that I use for food scraps is at the point where its contents need to be transferred into the compost container, prior to being ready to be taken out to the compost heap in the garden.

In such a passage of experience, my complex neural network responds to something in respect of which further notice, and action must be taken. In that further action the whole panoply of my being-in-the-world must (or should) be deployed so that I respond in a way consistent with how I act as a flexible and cognitively explorative being who is not only in the world but can think beyond what immediately presents itself. That empowers me in this world in which "I live and move and have my being" (as St Paul famously puts it—Acts 17.28), a human being lives a unified and integrated life as a result of his or her neurocognitive nature. It is a gross simplification of this type of being to reduce it to a denumerable set of functions able to be individuated and then joined up into some functional model of the human mind and that mind's way of thinking (suitably gender-neutral) because that mind is open to other worlds and their own counterfactuals.

In light of these current and future realities of life, it is important to note how Wittgenstein did transition from a logico-mathematical model in all but the last part of the *Tractatus* to a more socially and culturally complex theory of meaning and understanding, with all their indetermina-cies and intractability, in his later work. The puzzling conceptual location formed by, *inter alia*, 'the investigations', makes it understandable that some retreat into scepticism and failure to embrace the whole haunted some groups of later commentators. Yet that whole enjoins us to see that a different frame (or intellectual context), perhaps more sympathetic to qualitative analysis, will provide a way forward in the reflective endeavour. Framing theory in cognition and decision analysis allows one to mathe-matically assign certain values to biases and points of salience or impor-tance when our decision-making subject is mathematically modelled, and the provenance gives away the genealogy of the psychological theory involved. That genealogy is deeply tied to the economic model of human-ity which is an intellectual product of late capitalism. As such, it is a suit-able target for post-modern critique. Proceeding in this way brings us closer in some respects to Michel Foucault (1970). One of us has noted that convergence between psychology and economic theory in a reductive quasi-mathematical direction before (Gillett, 2008, 2009) but it seems particularly important at the present time. It is instructive to combine it with the debate between 'individualists' and 'communitarians' about rule following.

That debate is cast as a simple contrast between understanding linguis-tic meaning as a function of group activity *versus* understanding 'meaning' as it relates to the life of an individual in the world. The trick here is to do as Wittgenstein did and to transcend the private individual. This transcen-dence is not in terms of a set of internally organised private individuals treated statistically or quantitatively but of an evolved and highly devel-oped animal in a community engaged with contingencies and sensorimo-tor encounters. The public sphere is a sphere of adaptive activity under pressures of environmental dysfunction or flourishing. Both are meat and drink to an Aristotelian analysis that properly considers everyday life and the multifaceted ethology of human adaptation as a successful apex spe-cies, with all the complex interactions and interconnections between organisms and the environment. As the cubs say: "DYB DYB DYB; DOB, DOB, DOB" (Do Your Best … Do Our Best). This chant derives from 'the jungle books' where Rudyard Kipling attributes to the wolves the doggerel "The wolf is the strength of the pack, and the pack is the strength of the wolf".

Like wolves, human beings are 'beings in the world' surrounded by 'facticity' contingencies as they exist in an ecosphere. Within this framework, existentialism acknowledges human autopoiesis, a dynamic problem for mechanistic theories of mind and personality. We ought to note that human beings interact with the world by creatively and collectively developing techniques and sharing them through language. This is human 'self-making'(*auto-poiesis*). "In each case there is a reality—*Dasein/Ding an sich/tuche* that is signified so as to enable the mind to grapple with it in a meaningful way. But the description, designation, or intention or object cannot displace the actuality of our understanding of the world as embodied actors within it" (Gillett, 2008, p. 102).

It is also useful to begin to relate Wittgenstein's analysis of the complex human interaction with the world to the vexed problem of consciousness and its embodied realisation (Gillett, 2008, p. 100). This emerges particularly powerfully in his late work *Culture and Value* (1998). Here he links many elusive and dynamic aspects of our rich relationships and creations to our ways of being as embodied creatures. He remarks, for instance, on "the delightful way the various parts of the body differ in temperature" (1998, p. 11). One might not even notice or be conscious of this feature of being human until something like Wittgenstein's remark draws attention to it. And this is but one example of how our ways of being are not just conscious but also unconscious. In addition to temperature and other biological properties of the body, our sensorimotor functions occur, in part, outside our awareness but are critical for our ability to adapt to changing natural and social circumstances.

In becoming conscious of something, we bring it into salient presence and the holistic infinite possibilities here make us conscious simpliciter. It is no wonder that this complexity, unable to be rendered as an abstract thing or property of a scientific kind, should seem so mysterious and elusive. One might even be tempted to speculate about 'quantum reality' here.

All this grounds human beings in a spatiotemporal context in which certain objects remain relatively constant in position and character. This character is in dynamic flux with human faculties both sensory and motor which allow us to neuro-cognitively characterise them. We capture this complex and growing cognitive domain, as we have noted, by the grammatical device of 'objects' and 'predicates,' or 'denotation' and 'description'.

Neurodynamics, as applied to an organism located in an ecological niche, is a more naturalistically acceptable way to understand the psyche or animal mind than traditional Anglo-American thinking about evolved creatures understood as being a causally driven, genetically determined, biological machines. This anomaly of post-industrial thought has until recently infected both philosophy in general and neuro-philosophy in particular.

Analytic philosophy often associates 'being-in-the-world' with Continental and more fluid, or less precise and scientific, approaches to the human psyche and the whole panoply of psychological function. It is worth bearing in mind that these approaches tend to be rendered in terms more familiar to those inclined toward Anglo-American philosophy of science, the presumed cutting edge of philosophical naturalism. In truth, the failure of the Enlightenment project and the importance of language as a dynamic and transformative epigenetic influence on human beings was recognized by Wittgenstein as his work emerged from the Fregean and Russellian framework into which the naturalistic program in analytic philosophy had transformed itself.

Wittgenstein realised that ways of going on in the holistic and living context of human life would not submit to because they could not be captured by a quasi-mathematical analysis. This thought surfaces in his *Philosophical Grammar*. Language begins to live and have meaning as part of human life fully understood.

> Did you mean that seriously or as a joke? (#3)

> The different experiences I have when I see a picture first one way and then another are comparable to the experience I have when I read a sentence with understanding and without understanding. (#4)

A very different 'picture theory of meaning' from the early Wittgenstein emerges. It is not propositions, as logical pictures of facts or states of affairs, but instead life and practice that shape the meaning and our understanding of experience.

"I can use the word 'yellow'" is like "I know how to move the king in chess".

And so, by ushering onto the stage the interplay between skill and rule-based knowledge in the game of chess we become readied for the idea of 'language games' (*Philosophical Investigations*, pp. 23ff).

As Wittgenstein pointed out, the game of life consists not just of language but a broad range of behaviours. It includes socio-political influences on these behaviours and results in a performance designed to present the person in a certain light or framed in a certain way to the intended audience. Each of us has a number of different personae which can be exhibited on different occasions depending on the context (or frame) of the portrait so presented. These are different manifestations of the soul (*psuche*). They may involve wishing to be a successful academic, a person of the world, a 'man for all seasons', the man or woman for a crisis (keeping your head when all around you are losing theirs), the empathic friend, or someone able to discern a clear way to resolve a problem when confusion and stultifying complexity rules the mood of some moment.

All these scenarios are variously familiar to us all from various moments in our lives. We have either learned to negotiate them or endured a period of life-crisis, as can be caused by a stroke or head injury, mental illness, cancer, or other disease. Yet somehow the majority of us pull through, increasingly many as we adopt rehabilitation paths which engage with the wider human world. Tragically, some do not re-emerge into or re-engage with the world. The harrowing suicide statistics particularly in developed societies, with many intersecting concerns confusing and distressing people at times of challenge and transition, bear testimony to the casualties of the human need for flexibility in the face of adaptive pressure from sources external to us. We discussed some of these casualties from various neuropathologies in Chap. 2. For those who lose this flexibility and the cognitive and affective capacity to adapt to environmental demands, rehabilitation paths must be provided for us by interventions provided by social and medical institutions that can help to restore this capacity.

CHANGING FRAMES

The frames available to any speaker or thinker can be thought of as analogous to the spatiotemporally indexed referents within these frames. They are sites of action and perception and social contexts available to that thinker as a result of their unique experience. These provide points of entry into the interwoven and holistic neural network linking human function, relationships, and styles of interacting that are attuned to the rhythms of life. That is very important in every culture, and many facets of a person are signalled in a personal proper name, which has varying genealogy in different cultures. This is both idiosyncratic and autobiographical. When we consider the frames available to a human individual in the sense of contexts against which new ways of going on might be developed in the neural net, we turn to the person's biography as lived from a first-person perspective.

A biography of being-in-the-world from this perspective is an autobiography formed from one's actions in the face of contingencies (Jaspers, 1938/1971; Sartre, 1945/2007). This 'text' affirms the primary role of the person as an individual with a coherent and integrated mental life. The account is a cognitive attainment and not merely a mental state whose phenomenology and content are divorced from the external world. Changes in one's autobiography from one stage to the next can alter its contours in salutary or deleterious ways. This may depend on the level of functioning and flourishing one had before these changes occurred. The second type of change is discernible when we consider cases in which brain damage works its disruptive influence on a person's ability to negotiate life changes, particularly when, until the crisis happens, life has been 'sweet'. Such is the case with gifted young athletes, particularly those in professional sport. Such individuals are usually of high status and hugely supported both emotionally and financially. Closed head injury, with its highly disruptive effect on cognition and psychological adaptability, (often arising within the sport at which they have been so successful), throws all into disarray. A personal and moral crisis ensues which is not only moral but also psychological and potentially mortal (Gillett & Butler, 2021).

In some cases of young athletes, the crisis may include the realization that "the name died before the man" (Housman, 1896/1995). The narrative may consist of bright early chapters and a long, dark denouement. If there is to be a road to recovery, it is difficult and requires a fellowship (as in 'the Fellowship of the Ring'). Such a fellowship pulls one back from 'the gates of Mordor' and the dangers therein or restores one to the world after the journey.

The recovery journey is difficult but inspiring, indeed a testament to the human spirit. The narrative continuity of a life, and the need to be inclusive of all the interactions and rhythms of life that the person has developed, adds a significance to flexible being. This can be summarised in song; one gets there "with a little help from my friends" as Ringo Starr tells us. The help is in attempts to restore functions or enable the individual to adapt to physical changes. It may also be needed in attempts to enable people to adapt to radical cultural changes, as explained by Wexler. Whether one does or does nor adapt may depend on having psychological resources like Luria's patient Zazetsky.

Another syllogism can make explicit the difference between the dynamic contextual approach to how we think and act and a functional psychology of states and events.

2.1 Dynamism vs Function
1. The person in the world lives according to 'rhythms of life'.
2. The rhythms are dynamic and responsive to longitudinal complexity.
3. States and events are abstractions from that dynamic interaction.
Therefore
4. The psychology of the individual is dynamic rather than comprising codifiable states and events.

This argument refutes the claim that mental states and events abstracted from the subjects who have and use them in their actions can fulfil the requirements of a functional scheme and causal account of human behaviour. The abstraction involved in that scheme cannot do justice to human neuropsychology naturalistically conceived. The idea of mental terms, absent any metaphysical or scientifically recognised role for the person as an embodied and embedded moral and relationally identifiable individual, is palpably flawed.

The conscious person in the fullest sense of a being engaged with the natural and social world is common to almost all indigenous accounts based in families, tribes, places of origin, and sites of development and maturation. The resulting view of the person is radically at odds with the disembodied and disengaged cognitive abstraction of mind and meaning in the pseudoscientific explanations of them of the post-industrial model.

Chaos and complexity are features of the entanglement of a human neural net in the world. Because of the dyadic and ethologically situated relationship between this net and the world, myriad non-specifiable conditions open in a two-way exchange between a human being and their natural and social surroundings. These conditions open for adaptive change under higher direction and focus from the brain in enabling flexibility and cognitively mediated change. This 'higher' direction and focus does not fall into 'natural kinds' of biological specification. Rather, it is inclusive of cultural and historical nuances that often inflict a so-called 'moment of experience' so that it will never be exactly repeated. The result is apparent chaos in the nervous system that, over a time period of milliseconds, resolves itself into a pattern of oscillating and synchronizing rhythms corresponding to changing life-situations and sensitive to its autobiographical familiarity.

This mode of epigenetic evolution is characteristic of all higher animals. When an animal encounters conditions where a way of responding that is frequent or well-practised meets novel negative consequences (as one of us saw with the experimental rats), the animal searches for ways to find itself at home in a familiar 'groove' with all the reassuring sensorimotor signals involved in that response. The whole ethos of that complex of dynamic events is beautifully captured by jazz compositions such as 'Take 5', or 'So What', and how these and other pieces have been influenced by The Blues. The reassuring signals are like the chords or rhythms in these compositions that provide the basic structure over which the musicians improvise according to their unique auditory—and indeed more general life-experience in interpreting and responding to it.

Human cognition is therefore intuitive rather than formal and structured. It is spatiotemporally located and culturally framed (Tomasello, 1999). This frame needs updating and adjustment depending on the cognitive demands of the environment on the subject. One is reminded of the bends in the River Thames—as distinct from a straightish river—when trying to find one's bearings in London. One eventually adjusts so that the curves and loops in the Thames come to form part of a cognitive framework within which one navigates one's way around London. Places like The City, The East End, and Knightsbridge, almost take on a rectilinear geography which, while a distortion of reality, are good enough to find your way about. A further complexity enters when one takes on and learns to transition from the idealisations in the underground system and its route maps to the reality of a walking experience that takes in one's favourite haunts. Thus, London itself, replete with novelty and familiarity, can be a poignant lesson about 'representation and reality' in a way that no other extended logical or linguistic fragment of experiential life could be.

SERENDIPITY AND AUTOBIOGRAPHY

The human cognitive map is shot through and through with serendipity and autobiographical whimsy. We variously enjoy such experiences as the fantasy world of 'Winnie the Pooh', or 'Narnia', or the semi-historical, Anglo-Saxon world of Bernard Cornwell's 'Uthred'. In fact, thoughts informed by Cornwell's Anglo-Saxon world evoke what Great Britain itself might have been like to live in during that historical period. That 'world' is revealed to us in a new way and allows us to visit an historically recreated picture of England and the other UK countries so as to shine fresh light on them. We narratively indwell places and recognise their names. We do this however unfamiliar those Anglo-Saxon names are to us. As we get to know an historical or credible character and trace their story through events real or imagined, we delight in the way that we see character formed in front of our readers gaze. We admire the way that chance and necessity are skilfully constructed or revealed for us so that the character with whom we are engaged becomes convincing and our minds are enthralled.

Although some aspects of our lives are limited by the necessity of natural laws, there are also contingencies that fall outside the domain of what we can predict. How we respond to them is a measure of our adaptability to changing circumstances. They provide opportunities to construct an autobiography, the unique imprint of our individual response to them. These are events for which we cannot plan and which we cannot anticipate. They force us to act spontaneously in the process of self-creation. Thus, much of each person's narrative is a product of serendipity. This is essentially Sartre's first principle of existentialism (1945/2007), where one does not conceive oneself to be but wills oneself to be in the absence of objective guideposts for action. Each subject makes oneself through one's own actions as a being in the world.

Armed with that framework for the understanding of human psychology and personality, a much more fertile soil for humanistic neuro-philosophical, cultural, and ethical thinking emerges. The new dynamic approach also articulates well with philosophical explorations in aesthetics and post-colonial thought.

Doppelgangers and Brain Transplants

Doppelgangers and brain transplants fare quite differently under the present framework for neurocognitive life. Brain transplants, whether partial or complete, have always looked hopelessly fanciful. Advanced animal neurocognitive systems are intricately interwoven with the world through ongoing rhythms of sensorimotor interaction, as Hughlings Jackson realised (Franz & Gillett, 2011). Cognition supervenes on this dynamic system. It is woven into the evolutionary history and lived reality of animal life as a constantly changing dynamic system. Cognition updates itself according to subtle to-and-fro micro- and macro-alterations involving embodied multi-level sensorimotor feed-forward and feed-back changes linking cognition to *somaton* generally and holistically. Thus, brain transplants are likely to violate a number of difficult to abstract metaphysical constraints on human psychological life as essentially shaped by the natural and social milieu.

Nevertheless, they reinforce a point we made in Chap. 2. The idea that the same mental states can be maintained in moving a brain from one body to another is implausible because each person's brain is uniquely embodied. Similar remarks apply to the idea of a brain in a vat, purportedly active within a self-contained artificial medium (Putnam, 1981). The realistic ecological view supports the idea that psychological life depends on interaction between a subject's brain and the environment as well as how one interacts with the world as an embodied and embedded subject.

A syllogism is again apposite.

2.2 'Brains' vs embodied neurocognitive systems

1. The brain contains all the important neuro-cognitive 'stuff' of a person.

2. An organ can be transplanted into another body.

3. The neurocognitive basis of a person is transferable between bodies.

But if we substitute the realistic

1a. The brain *in situ* in the situated body to which it is dynamically in relation is the neurocognitive core of a person.

We arrive at:

3a. The situated embodied brain is the neurocognitive basis of a person and cannot be parted from its proper body without loss.

Doppelgängers are in a different conceptual context. These are imaginative point-by-point reproductions of a human being replete with the inscribed rhythms of rules and somatic ways of going on and adapted to an ecological context. In many respects, they can take off from where the real person left off (if 'left off' they did). Recognition of people and places, slotting into contexts, awakening, and using prefigured rhythms of life and interaction, all would be possible for these entities. Of course, from the point of the *doppelgänger* taking over, new complexities of association would start to form so that a gradual departure would proceed away from the person who is historically real to the one who is holistically self-making in a lived world of being and becoming. The result will be a subtly different person, just as inflected by situated socio-political reality as the original but with a slightly different outcome depending on lived time. But then, in fairy tales, we are not doing naturalistic metaphysics (Wilkes, 1994).

CONCLUSION

The sense and reference of the words we use to express ourselves and communicate with others cannot be analysed apart from the natural and social contexts in which we use them. Their meaning cannot be explained in functional or mechanistic terms devoid of the cognitive and affective capacities and properties of subjects engaged in the process of understanding each other. Instead, it is explained by how we use the words, and how they are shaped by spatial and temporal features of the different environments in which we live and act. Meaning and understanding are products of linguistic dynamics that connect speaker and hearer in ecologically situated speech acts in cultural interactions.

Wittgenstein's realisation of this phenomenon and his extensive discussion of it in his later work allowed the philosophy of language to evolve beyond its logical and mathematical foundations in Frege, Russell, and the early Wittgenstein to an appreciation of its essentially social and cultural role in a broad range of human behaviours. Our use of language in different contexts is one aspect of flexible and adaptive agency in responding to the cognitive and affective demands that arise within them. We use language, not just to *assert* things, but to *do* things (Austin, 1962; Searle, 1969; Fogal et al., 2018).

The brain from which the mind emerges is not a self-contained and isolated entity but a relational organ that connects us to and enables us to engage with the world (Fuchs, 2018). Our brains and minds are uniquely embodied and embedded, spatially, temporally, socially, and culturally. This is what makes each of us a distinct person and captures the true sense of being in the world. The dynamic neural rhythms that are shaped by these contexts form the neurocognitive basis of our behaviour. We are now in a position to go on and consider what it means to behave flexibly in terms of science and human nature. This is the focus of Chap. 4.

REFERENCES

Austin, J. L. (1962). *How to do things with words.* Clarendon Press.

Barwise, J., & Perry, J. (1984). *Situations and attitudes.* MIT Press.

Davidson, D. (1969). True to the facts. *Journal of Philosophy, 66,* 748–764.

Eliot, T. S. (1915/2002). *The love song of J. Alfred Prufrock and other works.* Amereon.

Evans, G. (1982). *The varieties of reference* (J. McDowell, Ed.). Clarendon Press.

Fodor, J. (1987). *Psychosemantics: The problem of meaning in the philosophy of mind.* MIT Press.

Fogal, D., Harris, D., & Moss, M. (Eds.). (2018). *New work on speech acts.* Oxford University Press.

Foucault, M. (1970). *The order of things: An archaeology of the human sciences.* Routledge.

Franz, E., & Gillett, G. (2011). John Hughlings Jackson's evolutionary neurology: A unifying framework for cognitive neuroscience. *Brain, 134,* 114–120.

Frege, G. (1892–1918/1980). *Translations from the philosophical writings of Gottlob Frege* (P. Geach & M. Black, Trans. and Eds.). 3rd ed. Blackwell.

Fuchs, T. (2018). *Ecology of the brain: The phenomenology and biology of the embodied mind.* Oxford University Press.

Gallagher, S. (2020). *Action and interaction.* Oxford University Press.

Gillett, G. (1992). *Representation, meaning and thought.* Clarendon Press.

Gillett, G. (2008). *Subjectivity and being somebody: Human identity and neuroethics.* Imprint Academic.

Gillett, G. (2009). *The mind and its discontents: An essay in discursive psychiatry* (2nd ed.). Oxford University Press.

Gillett, G., & Butler, M. (2021). When the music's over, then dancing with a partner will help you find the beat. *Cambridge Quarterly of Healthcare Ethics, 30,* 631–636.

Han, S., & Northoff, G. (2008). Culture-sensitive neural substrates of human cognition: Transcultural neuroimaging approach. *Nature Reviews Neuroscience, 9,* 646–654.

Harre, R. (1984). *Personal being.* Harvard University Press.

Housman, A. E. (1896/1995). To an athlete dying young. In *A.E. Housman: Collected poems.* Penguin.

Hughlings Jackson, J. (1887). Remarks on the evolution and dissolution of the nervous system. *British Journal of Psychiatry, 33,* 25–48.

Jaspers, K. (1938/1971). *Philosophy of existence* (R. Grabau, Trans.). University of Pennsylvania Press.

Kripke, S. (1980). *Naming and necessity.* Harvard University Press.

Lorenz, K. (1952). *King Solomon's ring: New light on animal ways* (M. Kerr Wilson, Trans.). Methuen.

Lorenz, K. (1966). *On aggression* (M. Kerr Wilson, Trans.). Methuen.

Luria, A. R. (1973). *The working brain: An introduction to neuropsychology,* trans. B. Haigh. Harvard University Press.

Neale, S. (2001). *Facing facts.* Clarendon Press.

Neisser, U. (1967). *Cognitive psychology.* Prentice-Hall.

Neisser, U. (1976). *Cognition and reality: Principles and implications of cognitive psychology.* Freeman.

Northoff, G. (2014). *Unlocking the brain: Volume 2—Consciousness.* Oxford University Press.

Northoff, G. (2018). *The spontaneous brain: From the mind-body to the world-brain problem.* MIT Press.

Northoff, G. (2021). Why is there sentience? A temporo-spatial approach to consciousness. *Journal of Consciousness Studies, 28*, 67–82.

Northoff, G. (2016). *Neuro-philosophy and the healthy mind: Learning from the unwell brain.* W.W. Norton & Company.

Northoff, G., & Hayes, D. (2011). Is our self nothing but reward? *Biological Psychiatry, 69*, 1019–1025.

Piaget, J. (1929). The child's conception of the world. *Journal of Philosophical Studies, 4*, 422–424.

Piaget, J. (1974). *Understanding causality.* Norton.

Piaget, J. (1999). *The psychology of intelligence.* Routledge.

Putnam, H. (1981). *Reason, truth, and history.* Cambridge University Press.

Russell, B. (1905). On denoting. *Mind, 14*, 873–887.

Russell, B. (1912). *The problems of philosophy.* Williams and Norgate.

Sartre, J.-P. (1945/2007). *Existentialism is a humanism* (C. Macomber, Trans.). Yale University Press.

Searle, J. (1969). *Speech acts: An essay in the philosophy of language.* Cambridge University Press.

Shotter, J. (1984). *Social accountability and selfhood.* Blackwell.

Shotter, J. (1996). 'Now I can go on:' Wittgenstein on embodied embeddedness in the 'Hurly-Burly' of life. *Human Studies, 19*, 385–407.

Sinha, A. (2021). *Sraffa and Wittgenstein.* Springer.

Sissons, J. (2006). Domestication and historicity: In memory of Peter J. Wilson. *SITES: New Series, 3*, 137–149.

Tomasello, M. (1999). *The cultural origins of human cognition.* Harvard University Press.

Vygotsky, L. (1978). *Mind in society: Development of higher psychological processes.* Harvard University Press.

Wilkes, K. (1994). *Real people: Personal identity without thought experiments.* Clarendon Press.

Wilson, P. (1980). *Man, the promising primate: The conditions of human evolution.* Yale University Press.

Wilson, P. (1988). *The domestication of the human species.* Yale University Press.

Wittgenstein, L. (1921/1974). *Tractatus Logico-Philosophicus* (D. Pears & B. McGuinness, Trans.). Routledge.

Wittgenstein, L. (1983). *Remarks on the foundations of mathematics* (G. E. M. Anscombe, Trans.; G. H. von Wright & R. Rhees, Eds.). Revised edition. MIT Press.

Wittgenstein, L. (1991). *Philosophical grammar* (A. Kenney, Trans.; R. Rhees, Ed.). Blackwell.

Wittgenstein, L. (1998). *Culture and value* (G. H. von Wright, Ed.). Revised edition. Blackwell.

Woodward, J. (2021). *Causation with a human face: Normative theory and descriptive psychology.* Oxford University Press.

CHAPTER 4

The Neurophilosophy of Flexible Being

In the last chapter, we considered how Wittgenstein's later view of our ways of going on amidst the complex rhythms of life moves beyond Frege's, Russell's, and his own earlier conception of philosophical logic and language. It is time to turn our attention to what is often presumed to be rational cognition in its purest form: the syllogism. A syllogism begins with a true claim, albeit abstract. This abstraction is supposed to present the truth in a certain way, and we used some examples in Chap. 3.

An example drawn from health care ethics might be as follows:

1. Human life is made good by pleasures and bad by pains.

That is then followed by another true (enough) claim:

2. A life characterised by only pain with an imminent prospect of death should be allowed to end.
3. The person him or herself is the best arbiter of the worth of his/her life.
4. Human beings should be allowed to kill themselves.

Note that this argument is framed in general terms with little in the way of 'the smell of the clinic'.

G. Gillett, W. Glannon, *The Neurodynamic Soul*, New Directions in Philosophy and Cognitive Science,
https://doi.org/10.1007/978-3-031-44951-2_4

Against this, one might embark on a different abstract thought on the subject.

2.1. People are sometimes mistaken about the value of their own lives.
2.2. That mistake might lead them to do something precipitate or foolish.
2.3. That precipitate or foolish thing might end their life.
2.4. A person should not be able to take their own life.

This argument, at first glance, seems to 'smell' a little more of clinical reality. Indeed, a recent review of pain management suggests that the abstractions divert attention from the real issues in many pain sufferers (Higgins et al., 2018). This review surveyed the complex relationship involved with its many biopsychosocial facets. It followed an earlier collection about the meanings of pain (van Rysewyck, 2016). That collection introduced multidisciplinary foci into what had become a complex clinical specialty with all the lack of breadth that can be entailed.

These two opposing arguments and their associated literatures, whatever their genealogy, each begin with an abstraction that captures a plausible aspect of the thought of a human being leading a complex life in a world where disease and other adverse factors touch frail incarnate and sometimes suffering people. In each case, opposing factors go unstated in the interest of drawing some favoured conclusion. But a moment's thought reveals that the initial premises, though plausible, are quite selective in the abstractions that they make and tend towards different intentions to act. Once that debate (perhaps internal to a compromised individual) is done, a complex and possibly troubled state of mind and its logic lead the person toward a certain conclusion and the actions it prefigures.

But it would be a naive healthcare practitioner who did not realise that emotions and images affecting the premises, and therefore the conclusion, are important in developing a health care plan which may then be pursued with quite strategic thinking. These factors are missing from most hedonistic starting premises in that human life is more than pleasure and pain. Inter alia, relationships and interdependences are important. These can be preserved through suffering and sometimes mean a great deal (as does the lack thereof) and can mitigate that suffering to the point of it being especially meaningful to those affected. More holistic human factors can also be important in sustaining people through troubles that would otherwise be overwhelming. An accomplished academic with acute neurological impairment might, for instance, contemplate ending it all. That may, on reflection, be favoured by a deep-seated anxiety about losing neurocognitive powers and temporary dislocation from loving relationships. That is something that the presence of a sibling or a child and grandchild might reverse, restoring the effaced love of life.

ABSTRACTION AND NEUROCOGNITION

The cognitive, emotive, executive, and self-corrective features of our lives of thought and feeling must be alert to these complexities of embodiment and to the difficult and interwoven lives of the soul that "are never dreamed of in some philosophies"(even those of 'Horatio'). To be alert to those things is to be a developed and flexible neurocognitive being able to think and rethink the twists and turns of *l'etre et le neant* when we find ourselves enmeshed in them. It is of course also possible to be crippled by 'over-thinking' (De Haan et al., 2015; Fuchs, 2011).

Flexible being and a neurodynamic perspective are hard to appreciate and analyse our thinking when we attempt to work within the framework of philosophical abstraction. This includes the many post-industrial forms of such abstraction, such as the clean logic of causal determinism:

> This is the thesis that a complete description of the world at some time T, in conjunction with a complete formation of the natural laws, entails every truth at about the physical state of the world at later times. (Glannon, 2015, p. 3)

It is the unholy alliance of this causal and other mechanistic conceptions of the world and our place in it that is "the fly in the ointment" here. Among other things, it tends to 'cool' the human warmth that sustains us. Freeman's dynamic conception of human neural function reopens the philosophical and ethical discussion, allowing it to transcend a mechanistic world view and yet remain clearly naturalistic. This inclusive re-orientation affects our conduct and our contemplation, making us more reasoned and reflective about life. At the same time, it makes us more alert to the subtleties of our dynamic being in which we are enmeshed along with those we care for. That is where love and human identity live (it even affects our pets, sometimes to their detriment in terms of pleasure and pain).

The dynamic approach to our cognitive and biopsychosocial lives does not make the soul's task of living in the light of reason and contemplation any easier. It is, in fact, a philosopher who typically values logical and linguistic neatness rather than goodness, who wants us to be able to distil life and thought to syllogisms and clearly specified abstract terms, sentences, and arguments. Against the dominance of this abstract view in philosophy, existentialists have formed a new synthesis of thought and life that points beyond formal logic and language with their neglect of the lived human realty and their scepticism about their ability to capture the deeper meaning of being human. This depth potentiates an understanding of being in the world. That synthesis is part of our intention, rethinking thought about 'thought' in light of the complexity and evolving dynamic change in the neural network of an ecologically engaged creature—a human being. It is no accident that certain existentialists—for example, Sartre—are also accomplished novelists. We contend that an understanding of human cognition as it is/in existence opens the way for complexity theory and a conceptual transition that moves away from mechanistic causality and towards the real, lived world.

The resulting naturalism underpins a sophisticated form of neuro-cognition that is engaged with the physiology of the organism and can accommodate biomedical and biopsychosocial styles of explanation. In fact, the holism involved in understanding the rhythms of the brain helps cognitively focused neuro-philosophers to take realistic account of diverse recent advances in neuropsychopharmacology. These include the use of ketamine in suicidal ideation (Abbar et al., 2022) and psilocybin for depression (Carhart-Harris et al., 2017; Smith & Sisti, 2021). They have also made us relatively open-minded about natural and placebo effects from 'folk medicine' and herbal remedies in mental and behavioural disorders (Benedetti, 2014, Ch. 6).

The idea of holistic and embodied neural rhythms that develop through life and retain a certain flexibility into adulthood is also relevant to the uncertain area of the gender identity debate. It is particularly relevant to the idea of interventions to solidify this identity at certain ages (Mikkola, 2009; Knox et al., 2019). Mari Mikkola injects and examines a number of conceptual problems about gender (Mikkola, 2009). These include cognitive and affective influences which unsettle straightforward biological determinations based on somatic or genetic binaries and fail to recognise socio-cultural and developmental complexities.

Self-image, family dynamics, cultural expectations and roles, relationships with peers, physical attainments and expectations, interpersonal attractions, and sexuality are but some of the issues that shape human beings in this area. Negotiating these issues may take skills and experience which many young people lack. Developing these skills requires an environment providing scientific, psychological, social, and cultural support. It requires a biopsychosocial orientation.

The issues hit a crunch point in sport. The concern with sport is that many well-grounded biological characteristics guiding 'norms' (such as who can legally play in which teams in a competition) and 'records' (e.g., in Olympic weightlifting) in sport are susceptible to problematisation by socio-political changes. The problem has no easy solution philosophically, ethically, or politically. Neuro-philosophically sophisticated ethicists involved with young people and the clinicians associated with them, aware of the complex interacting rhythms of life, are pulled in many directions. Gender associated physiognomy may incline them to think one way, and yet neuro-cognitive changes controlling sensorimotor functions that are autonomic and not completely conscious, social life, expectations, and conditioning regimes all reflect or add to its complexity. These complexities arising from the interplay of biology and sociocultural effects again arise both in humans and the non-human animals we relate to as pets (house-training a cat for example). Higher levels of neural function, such as social and cultural life, affect widespread brute ways of behaving by exploiting neuro-cognitive plasticity.

Neurophilosophy and Culture

Our work may be considered to be in the tradition of Oxford scholarship, which Paul Snowdon examines in relation to P.F. Strawson. Snowden argues that Strawson was particularly to be esteemed for his ability "to stimulate thought" and do it in both "descriptive" and "revisionary" ways (Snowdon, 2008). The present work, hugely influenced by both Strawson and Snowdon, also aims to be both revisionary and descriptive, particularly in its treatment of consciousness.

The complexity and the role of consciousness in human thought leads us to ask 'what it is like' to be a person, a question made famous by Thomas Nagel (1974). As we have noted, young people especially are often struggling with sexuality, identity, and self-formation. We need to be particularly aware of the human and holistic process of forming an adult self of one's own. The thinkers engaging with this area foreground features of human neurocognitive life that are difficult to render in functional or operational terms. It requires sensitive *Menschenkenntnis* to properly explore these features.

One of us has examined the thought experiment of an individual called Mary who undergoes a 'brain rehabilitation'. That intervention disrupts a dysfunctional pattern of behaviour that has many of the characteristics of a profound depression. The question arises as to whether it has changed the real or authentic Mary as a psychological individual. Under philosophical examination, this difficult question seems to turn on a combination of narrative integrity and the holistic features of a naturally arising human memory, which is intricately interwoven with the rhythms of life (Gillett, 1985, 1988). Those rhythms interactively enable the brain to configure itself adaptively to its ecological niche. That radical bypassing of reductive metaphors used to operationalise psychology and human function as a creature in the world is meat and drink to a study of ethological neurodynamics and the rhythms of life as the neural basis of the human psyche.

Brain interventions of various kinds disrupt a person's life in various ways. Sometimes apparently minor impairments such as closed head injury or concussion are, in fact, devastating for the person concerned (Gillett, 2018). The effects and their subsequent impact on recovery are, on occasion, cruel, sometimes so as to cause either suicide or death by chronic alcohol abuse. This is a greater evil because of neglect or disinformation generated by entrenched interests such as the big business of commerce in sport or social expectations, such as always being good for 'a drink at the pub'.

A neurodynamic approach of an inclusive kind embraces the complexity of our interconnected ways of going on in the real world as cognitively flexible beings. This complexity involves heterogeneity in the type and extent of our cognitive capacities. As flexible beings, we are able to adaptively manipulate and explore the real world and imagine alternatives that may eventuate. Such alternatives are well-developed in post-industrial settings and their health care systems. They should also be sensitively potentiated and enabled particularly in post-colonial contexts where indigenous communities must become more recognised and respected so as to maximise our shared human adaptability. They are not only open to the contemporary world and its modes of life but also retain traditional mores at the centre of their being and self-worth. This may involve mysteries with which they resonate, and which cannot be explained away by the materialistic analyses of persons from alien or colonial origins. Recovering that resonance makes diverse individuals less likely to lose their integrity and cultural heritage and creates vibrancy and diversity in the communities in which they are valued.

Well-integrated thinking about the world so that it incorporates forms in which all cultures can feel they are coping with life and its many contingencies preserves 'culture and value' (to again resonate with Wittgenstein). This enriches us all, as many in New Zealand already have experienced. Of course, over the eons of time that indigenous ways of going on have been formed in an ecological context, their ways may seem very settled and unchanging, obeying as they do an ancient and ecological rather than a post-industrial timescale of change. The issue of climate change is, however, making many of us more aware of our short-sightedness, limitations, and conceptual deficiencies in a fragile world.

One of us has stated that the facts involved in flexible being as a person in the world are no less a fact (or facet) of life because they cannot be rendered in scientific terms. On that account, they join such things as tables, chairs, frogs—with all their cultural meanings, paintings, or questions such as who shot Claude's brother at the Battle of Waterloo—a fact if ever there was one (Gillett, 2009, p. 47). The facts involved force us to reconceive the essential nature of adequate neuro-philosophical accounts of human psychology and our thought about the real world.

Neurophilosophical accounts taking notice of the neurodynamic function of human cognitively moderated adaptation cannot be rendered in terms of operationally articulated strategies suitable for artificial intelligence as classically or currently conceived. The neural rhythms underpinning our sensorimotor and cognitive interactions with the world are always holistic and fluid in the face of experience and imagination. Whereas this fluidity may be simulated by an intelligent learning program, the engaged biological self, configured by the rhythms of life, is something unique. Its behaviour cannot be predicted or completely explained by such a program because how this self responds to contingencies in the world cannot be known before it responds to them. Contingency and the fluid rhythms of life form a complex res extensa that precludes predictability. It is part of the animal kingdom 'in the wild' and not of the domain of human artifice. But in the human case it is supplemented by counterfactual thinking about how to act when facing alternative possible courses of action in the 'garden of forking paths' extending from the present into the future.

A Holistic Approach to Consciousness

The neurophilosophy of the soul needs to embrace the many philosophical complexities that resist reductive analysis. It takes us into what could be thought of as a Thomist form of naturalism That philosophical reopening of the subject needs to happen, for one thing, in view of the biopsychosocial model of health and the post-colonial expansion of neurophilosophical thought beyond industrial metaphors (Bolton & Gillett, 2019; Gillett, 2016a). We therefore need a more adequate conception of the scientific study of processes in nature and our place in it. It is timely that this has become possible with dynamic, real-time neuroimaging, a primarily medical innovation, making it a fitting source for the neural grounding of biomedical philosophy of health and disease. As we have argued, though, functional imaging displaying glucose metabolism and blood-oxygenation levels needs to be supplemented by behavioural observation to explain how and understand why we act as we do. "An inner process stands in need of outward criteria" (Wittgenstein, 1953, #580).

This neurophilosophical repositioning of 'being human' leads to a revisiting of Nagel's 'what it is like to be' a person. Nagel spoke about 'facts' but, for the moment, we will suspend that characterisation, preserving only the intuition that subjectivity is a robust part of human psychology and another reason why that subject should not be understood purely in terms of causality or function (as in 'functionalism' or 'hard facts'). Rather, subjectivity transcends that limited conceptualisation towards fluid and living facts. These are woven seamlessly into what we will follow Māori Marsden in naming it as 'the woven universe'. For that reason, again, throughout this discussion we embrace a more holistic and inclusive approach to the human *psyche* including often neglected features stressed by Continental and aesthetic thinkers (Kant, 1790/2007; Schiller, 1795/2016).

Such thinkers have forged a possible rapprochement with a neuroimaging-based scientific body of work that grounds neurophilosophy firmly in empirical evidence (Bayne, 2009). That evidence is closely related to and correlated with real healthcare in such a way as to take the project forward in relation to the rhythms of everyday human life as lived. Tim Bayne, for instance, weaves the phenomenal content of human cognition (an enduring puzzle for post-industrial neurophilosophers) into the holistic top-down configuration of the human mind and its neurocognitive exposition. He argues that subjectivity (replete with qualia) introduces a sense of the human experience as an integrated and complex way of being, against the conceptual analyses of 'scientific' or more causal-functional psychology. In so doing he helps 'cut the Gordian knot' that many scientific thinkers (including quantum theorists) have woven around 'qualia'.

Bayne argues for what he calls 'the liberal view of perception' whereby the items, qualities, and events that are melded together to form our perceptual content are themselves interwoven in a way that invites a holistic analysis making use of complexity theory. Such an analysis reinstates *autopoiesis* and the human subject as an integrated whole into a prominent role in an adequate picture of human biopsychosocial being. He uses such telling examples as recognising the yellow of sunflowers or seeing a telephone as a telephone to introduce a fully lived life into perceptual psychology and does so in such a way as to completely undermine a bottom-up view of perception (Bayne, 2009, p. 387). He also takes the holism that can be applied to perception as having implications for modal debates which ground cognition in simple propositional elements or terms that can apply across a range of counterfactually conceived and constructed creatures. He is thus exempt from Wilkes' critique discussed in Chap. 3.

Bayne uses telling examples such as a non-French speaker or listener hearing the auditory stimulus 'il fait froid'. The experience is not apparent to anyone who only lists sound events as characterised by frequency and intensity. He compares such an experience with the characteristics of associative agnosia as described in the literature. In both cases, an aspect of the stimulus which transforms the whole would be missed by the perceiver. Seeing a tomato as a tomato, for instance, invokes gustatory, nutritional, and cultural aspects of the use of tomatoes in our diet, aspects which embrace the whole of our life as embodied beings in an ethological context (including its delightful culinary traditions). The approach outlined offers a synthesis that transcends the antithesis, vehemently debated in philosophy of science, between 'idealism'—of some form—and 'scientific realism'.

QBism (pronounced "cubism"—'Quantum Bayesianism') is an approach to metaphysics taking account of subjectivity, indeterminacy, quantum entanglement, belief, and action. It is interpreted by some as a kind of participatory realism. That stance would be consistent with this work in which participation in a real world is the ground of our being. We might, however, suspend for the moment a commitment to Bayesianism in order to accommodate a more inclusive understanding of human life, cognition, and socio-political discourse.

Bayne himself ingeniously considers certain rejoinders from the 'bottom-up' materialistic thinkers whom he has attempted to refute. In these rejoinders, he reconsiders contributions from various aspects of conscious experience (albeit characterised as an information processing totality). In so doing, he awakens thoughts such as that of seeing a tomato as a tomato, or any reading of 'seeing as' which introduces the totality of our being in the world. That includes our nourishing ourselves as creatures, which aspect of being human is replete with culture and 'flair'. Those qualitative aspects of life as it is lived enter into the appreciation of the experience of being human. Bayne modestly contends that the conservative materialistic and possibly reductive view of neurocognition as we analyse it could perhaps find a more accommodating analysis were we to allow that higher-level cognitive elements could enter into the total experience of certain phenomena. That makes any view that the rhythms of neurocognition resonate with the rhythms of life in an irreducible way seem much more attractive and opens the enquiry to our incarnate or protoplasmic interaction with the world.

The openness of the human life of the mind to both the well-known experiences of everyday life and the unknown mysteries of our dwelling within the ecosphere (the boundaries of which we do not fully understand) is evident in the religious thinking which many strands of humanity embrace and reflect upon. These include not only the mystical religions but also ways of thinking such as the *via negativa* in doctrinally inclined religions such as 'Eastern Orthodox Christianity'. Doctrinal versions of holy scripture-based religions are abstractions. As such, they are often replete with paradox as in The Bible with its recognition of the 'God' of the 'burning bush' of Moses and the 'still small voice' of Elijah. Such moments in sacred writings, if reflected upon, warn against the religious 'scientism' embraced by some as 'true and pure' doctrine as a basis for faith.

Consciousness and human responsibility as topics related to brain function continue to provoke controversy. The controversy seems to be fiercest when we begin to discuss their scientific basis, where, for some, the mysteries of subatomic physics have displaced dualism. Here, perhaps we need to recall Descartes "I think, therefore I am". We can easily overlook a crucial point: The claim concerns an indexical which is a discursive tool that has no place in science. Science, as Nagel reminds us, is "the view from nowhere" as it must be to focus on universal objective truths (Nagel, 1986). Thus, science can only teach us limited things about the neurophilosophy of consciousness (if it is known indexically).

In the *Philosophical Investigations*, Wittgenstein engages us with this issue as a philosophical truth which he relates to 'grammar':

> Human beings ... are their own witnesses that they have *consciousness*. ... Why not simply say "I perceive that I am conscious"? ... why not say "I am conscious"?—But don't the words "I perceive" here shew that I am attending to my consciousness?—which is ordinarily not the case. (1953, 416)

He examines and reflects upon situations where such locutions as "I am conscious again" would have a use. He notes, for instance, a moment in which I tell a doctor "Now I am conscious" (after, for instance, a concussion in sport). In such situations, Wittgenstein acknowledges that the words have a use and therefore are meaningful. Once again, we see language returned to its natural setting: real human life where words *do* something.

We must therefore approach the neurophilosophical debate about consciousness with a healthy suspension of belief or resistance to total immersion in Metaphysics. The upper-case 'M' is a precept/device equally applicable to religion when it comes to over-thinking physical rituals or metaphysical details of belief. This resistance is justified when philosophical models abstract from our everyday use of words and the actions that accompany them and are properly engaged with life.

Changeux and Dehaene's global neuronal workspace model is a prominent exposition of the present orientation of consciousness embodied in organisms embedded in the environment (Mashour et al., 2020). It includes a holistic and interactive model of the human organism as neuro-cognitively engaged in being in the world and generating adaptive ways of going on. Philosophically, however, the present analysis has a crucial element which, as noted, is indexical and therefore not 'scientific', because science deals only with general 'view from nowhere' and linguistic terms able to be observed objectively and not replete with qualia (note the 'grammatical' nature of the argument).

Despite the artistic licence in its use of science, we can also draw a lesson from the title of a recent film, "Everything All at Once Everywhere" (Abbar et al., 2022). The brain in the person who is 'a being-in-the-world' interactively and dynamically engages with others in multiple ways according to our many networks of concepts. These are able to be variously cognised using the associated languages: of science, aesthetics, engineering, mathematics, and others. The existentialists constantly remind us that representation is not reality as we live it and as it sustains our being in ways continuous with the lives of microbes, plants, and animals. In distinction from other species, we have 'grammatical' language which can be metaphorically used to convey things about other life-forms. Con-scio-ness ('knowing with') can be used of non-discursive creatures in the sense of 'know' which embraces both 'head' and 'heart' and even styles of doing things. For example: "She knows wood like you and I know the city we live in". These considerations confirm Chris Frith's point that "all the contents of consciousness are the outcome of a social endeavor" (Frith, 2008, p. 240). Emphasising this same idea, Adam Zeman states that the function of consciousness is to "free the organism from control by its immediate environment" by enabling flexible and adaptive behaviour within it. He adds that, as conscious subjects, we "are in and of the world from the start" (Zeman, 2008. P. 316).

A metaphor we might be tempted to use here is 'a rabbit hole' (with due nods towards Lewis Carroll, embroiled as he was in the academic 'pressure cooker' of Oxford). 'Carroll' (or Dodgson) must have been very familiar with the creative fervour of scientific thinking and its philosophical implications because of being in an 'Oxbridge' college with many internationally renowned voices. The 18 (or some other rarefied and taxonomized number) of fundamental particles or 'wavicles' in various scientific ontologies, or lists of 'existing things', and phenomena (including such difficult to grasp 'things' as 'gluons', 'strings', or the more cognitively accessible 'positrons') abound and jostle for epistemic and explanatory dominance. These are almost as exciting and varied as anything a 'ten feet tall' mind could devise and take with it on a 'Jefferson Airship'.

Here we might also consider a further syllogism:

3.1. Scientifically informed physics generates paradoxes.

3.2. Paradoxes indicate that their embedded claims are not strictly compatible.

3.3. Chaos and complexity theory and their embedded metaphors are complex ideas defeating a literal reading of syllogistic abstractions.

This syllogism warns us against thinking that the rarefied theories of complex sciences should be allowed to bypass the 'simple-minded' objection that representations represent and are not the reality being represented. This is often implicitly associated with the apparent romanticism of the existentialist philosophers and their apparent lack of scientific or logico-mathematical rigour (we might profitably recall Frege's 'grain of salt'.

It also seems a good fit for a naturalistic philosopher disenchanted with neo-Fregean or Russellian mathematical style of logicism (as in the *Tractatus*). In view of the indigestible paradoxes generated, the foolishness of attempting to 'swallow an elephant while straining at a gnat' comes to mind. That seems deeply apposite in relation to the complexity of cognition required to stay up to date with the conjectures and theoretical creations of modern physics. But this judgement is harsh and perhaps reflects one philosopher's characterisation of himself as 'a bear of small brain' when reading an essay by one of the authors.

CULTURE, CONTEXT, AND MEANING

The metaphysical problems of the quantum seem to be a virtual 'Gordian knot' of conceptual complexity attached to the world involving scientific endeavours. We can here draw a tenuous analogy with The Napoleonic Wars and the many possible explanations of British ascendency in battle. If we assess the contribution that can be made by the fact that the British army used rifles (Baker rifles) in some of their light companies and their riflemen's units and therefore often used their accuracy of fire to strip the French armies of officers, other facts remain significant. The importance of Baker rifles should neither be minimised nor overstated. The physics and ballistics of rifling gave the rifles distinct advantages, but the detailed physical ballistics assume a potentially over-stressed focus in some historical or even military discussions of the Napoleonic wars.

The Duke of Wellington's role in that nineteenth-century conflict was also an important factor. 'All decently and in order' one might urge and in so doing, strike the Aristotelian mean in discussing the British ascendance in the wars while recognising the complexity required of any such discussion. This included British reliance on Wellington's 'dregs of society', who, by nature, were streetfighters rather than 'upright citizens'. This was recognised by the duke in his reflections, along with his own crucial role as commander and his care for his men: "My God, I don't think it would have been done it if I had not been there". This is his famous comment on the victory at Waterloo. He also famously lamented the severe losses of men in the battle (in addition to clearly recognising their marginal psychosocial origins). The complexity of history reflects not only the many possible and actual recollections and reconstructions of those events but also the variety of points of view (each from somewhere rather than nowhere, involved in the facts.

Māori Marsden, as we have noted, is a thinker worth engaging with at this point. He was born in the north of New Zealand in 1924 and made a point of visiting the UK and Europe where he discerned a deep resonance between Māori thought (as an example of an indigenous tradition) and the most recent advances in Western science and philosophy.

> I took the word 'hirihiri' which in Maoridom means 'pure energy'. Here I recalled Einstein's concept of the real world behind the natural world as being comprised of 'rhythmical patterns of pure energy' and said. ... This was essentially the same concept. (His interlocutor) exclaimed "Do you mean to tell me that the Pakeha scientists (tohunga pakeha) have managed to rend the fabric (kahu) of the universe?" I said "Yes". (Marsden, 2003, p. xi)

So much for the limited nature of 'primitive' thought patterns.

Marsden was an exceptional mind, and identified as such by his tribal elders, and later theological educators. He serves to remind us that the neurophilosophy we are attempting to outline here, despite its contemporary guise, has deep cognitive and evolutionary roots in the human psyche. His inclusive vision and reflection therefore finds certain resonances in the thought of the Duke of Wellington (in terms of the role of unusual individuals and undervalued groups), however contrasting their lives and callings from most of us in any society.

Human 'being-in-the-world' actively conducts itself not only with *l'etre* but also with a recognition of and reflection about *le neant*—what is not (but might be). This factor in human cognition, explored by Sartre in his major 1958 work *Being and Nothingness*, opens philosophical enquiry to include not only dynamic systems and processes but also imagination and aesthetics in the form of all kinds of artistic development and future-oriented vision. These complexities can all too easily be avoided by analytic philosophers (particularly philosophers of science) or else reduced to logical discussions of how actual states of affairs might have been, abstracted from human agency and grounded experience. But the imagistic life of the mind does not obey the rules generated by such a diminished conception. That conception is confined to theoretical models which are composites developed from both abstracted experience and non-existent non-experience. They are constructed with the aid of already codified moments and passages of thought derived from that embodied experience that we actually enjoy and endure but not always with their actual and complex conditions of origin and contingencies in clear view. These models are distinct from the alternative possibilities we consider in projecting ourselves into the future and forming and executing plans of action.

The genuinely insightful thinker experiments in thought with a finely honed '*savoir faire*' about life and its lived contexts. Those realities and contingencies, interwoven with our existence as lived, ground thought and affect in that lived reality in a realistically permissive way but with caution.

These conceptual moves reinforce the idea that neurophilosophy is not, except in thought-experiments, about a machine or AI system that has been designed for some specifiable operation. Instead, neurophilosophy is about an open-ended embodied relation with the environment in which fluid techniques of behaving and perceiving emerge from those already developed in a protoplasmic organism occupying a particular ethological niche. Conceived properly, this occurs in the singular historical and prehistorical tracts of human genetic and epigenetic evolution which have formed us. Such an enquiry is and must be inclusive in transcending the limitations of current Anglo-American philosophy of science by introducing a *rapprochement* both with continental philosophy and indigenous thought. The corrective is long overdue and can be creatively explored by cutting-edge neuroscience to the advantage of both those engaged in philosophy of neuroscience and its development.

Some years ago, Gavin Ardley spoke of 'voluntary active phenomenalism' as a way of combining Kantianism with Thomistic thought. In this regard, he partly anticipated Wittgenstein (Ardley, 1968). The unifying thought is that there is an underlying reality in which we are engaged, and which transcends our many attempts to categorise and systematise it. Thus, we are always defeated when we become overly in thrall to any of those systems of philosophy which we devise. Ardley refers to cognitive structures as *enta rationis* or cognitive artefacts fashioned for the purpose of making the complexity and wholeness of nature as the ground of our being more tractable for cognition and less transcendent of human specifications. In doing so, neurophilosophy takes proper note of our complex associative system and its constant, dynamic, self-modification under the influence of experience. The resulting neurophilosophy becomes conceptually linked to the many streams of transcendent thought that human beings have adopted throughout history. This should have profound effects in our reflection on the soul (*psuche*).

The effects are marked, particularly in relation to discourse. That complex of phenomena becomes a living and relational medium of human development rather than a kind of calculus with a rigid grammatical structure allegedly understandable devoid of any wider context. When we turn our attention to discourse, we are plunged back into language games, forms of life, representation, meaning, and thought (Gillett, 1992) as the 'midwives' of freedom and resentment (Strawson, 1974), 'culture and value', and 'the woven universe'(Marsden, 2003). A respectful silence is potentiated which is nevertheless 'promise cramm'd' like the air rather than an artificial and abstract '*Conceptus logico-philosophicus*' (with apologies to Wittgenstein, 1921/1974, #7). The silences which this respect brings in its wake could perhaps be because of 'that which cannot be said'. As Wittgenstein went on to note, natural language cannot be adequately explored within an artificial structure without, we would say, dynamic and evolving holistic incursions from the wider context created by human life in the world. It must be considered as a matter of language games and forms of life rooted in and arising from our natural history and ultimately culminating in culture and value (Wittgenstein, 1998) and ethics.

In fact, Wittgenstein lectured both on ethics and free will. These are entangled topics. If we think of promising and the prominent role it had in philosophy at a certain point, an interesting anti-reductive argument comes into view.

1. Promises cause people to do and think things in some future situation.
2. The physical sounds or visual stimuli comprising the promise are not effective as extensionally causal events at that time.
3. The mode of action of a promise is not physico-causal.

Philosophical interest in promises from Hume and Kant to the present helps underscore the argument. Elizabeth Anscombe notes, "Hume had two theses about promises: one that a promise was 'naturally unintelligible', and the other that even if (*per impossibile*) it were naturally intelligible, it would not naturally give rise to an obligation" (Anscombe, 1978, p. 318).

The topic of discursive obligations is revisited by Scanlon, who relates the general problem to human involvement in social practices (which we have gathered under human epigenetic adaptations). We have thus urged an expanded anti-reductive naturalism (Scanlon, 1990). This argument could be couched differently by adverting to customs or cultures and the socio-political order. Such an inflection considers culture as something in which we are enmeshed but not able to be analysed in terms of 'natural kinds and their biological adaptations'. Such an epigenetic innovation, involves more than 'physical events' under causal laws. Cultural and political realities generate active forces at work in this 'human-all too human' reality we indwell, whatever form it takes on a given occasion. A promise, we might say, is surely a mental event but its subvenience conditions cannot be given in terms of purely physical antecedents as its ontology is other and does not merely embed physically described or neuroscientific events, kinds, and explanations.

What is more, the relevant experiment would be hard to do. Try to think of dynamic brain imaging that would chart the 'state' that arises from making a sincere and serious promise that was not merely part of something quite different—taking part in an experiment for instance. A promise in real life is likely to be holistically connected with the relationships and circumstances to which it applies. And that moral and procedural judgement is made in the light of the shared understanding that there are circumstances in which real people would not regard them as binding—a murderer at the door asking about a guest, for instance. No doubt some commonalities would exist with other brain states combining key elements able to be introspectively sketched but, in that each context of making a promise is holistic and interwoven with complex biopsychosocial situations. A neurocognitive 'skeleton' is all that is likely to show up in subtractive and contrastive dynamic neuroimaging. That will carry a partial and suggestive relation to the lived whole (as a skeleton carries to a 'lived in' and living body). Both 'skeletal' entities or conceptualisations are potentially informative but only partially so and require an educated eye to fill out the details.

The Breadth of Mental Life

Promising has reminded us of the ramifying context of common human ways of going on and their embeddedness in our ethological context. Here we might return to two topics which have periodically and persistently surfaced in the philosophy of mind and philosophical psychology. They were repeatedly and continually present in the literature of the 1980s when one of us wrote a doctoral thesis. The topics are entangled with consciousness and are (1) qualia and (2) the holism of mental life, and prominent philosophers involved included Thomas Nagel (1986), Kripke (1980), and Davidson (1982). The topics they were internally related to (among others) were 'rigid designation', 'natural kinds', 'possible worlds', and the irreducibility of 'mental states'. Each deserves attention if only for reasons of submitting important concerns in the history of philosophy and particularly neuro-philosophy of the neurodynamic analysis we have outlined.

In this enquiry, we can be guided by Hughlings Jackson's philosophical stance which he called 'concomitance'. It is an anti-reductive view that states of mind are realised in brain events but have their own effective and productive relations where explanation can be sought (Jacyna, 2011). He remarks "Scientific materialism is only materialistic as to what is material, the nervous system" (Hughlings Jackson, 1887, p. 45). This fine distinction drew on any philosophical sources from British analytic philosophy of science to French thinkers. But Hughlings Jackson made it his own in his hugely influential body of work that we would now call neurophilosophy (in addition to his ground-breaking neurology and psychiatry).

We can begin by returning to qualia and the case of pain in trying to understand the role of this and other qualitative experiences in human life. The case is repeatedly being made that pain plays an extensive role in life and that 'false pain' or even pain originating from unconscious conflict can be relatively indistinguishable from real pain in clinical practice. Yet this range of phenomena and the experience they embed may be better understood by adopting a different clinical approach (Gillett, 2016b). Such an approach forms the basis of the clinical treatment of 'hysteria' (or factitious disorders) and tends to the conclusion that many of the related and complex roles played by pain in our interwoven, interpersonal, and medical worlds may have to be reflectively cut in error in order to understand the biopsychosocial role of this human phenomenon. The same considerations do not apply to other animals, though they can also 'practice to deceive' (without such 'tangled webs'—mother birds drawing a predator away from the nest by feigning injury, for instance).

Far from being purely a so-called 'phenomenal state', pain and other subjective or even qualitative experiences such as the chilly and inhibiting atmosphere in a meeting are powerful human experiences albeit with subjective and somewhat difficult-to-describe features. It is just such analytically elusive and holistic features of being a flesh-and-blood human being in the world among others that often escape precise specification in functional terms. The over-systemisation here is symptomatic of the intentional stance involving a level of abstraction to explain behaviour in terms of mental properties (Dennett, 1987). This served clinicians badly by generating a pervasive misunderstanding of subjective experience. Similar indeterminacy and conceptual difficulties attend the apparently clear-cut and extensively analysed notion of rationality so beloved by a large cast of philosophers. In fact, this idea is so entrenched that some are prepared to hang the whole notion of thought or cognition on that foundation rendered in suitably abstract and tractable form.

Consider, for instance, the remark: "He seems a little out of sorts today". This remark is perfectly comprehensible, if less than precise. It could be part, for example, of a mental explanation of why a general might not be able to competently plan a crucial campaign, why a doctor might not conceive and justify a diagnosis and regimen of treatment, or why a cricketer might not think out and execute an innings given a particular situation in a test match. Each of these human situations requires a complex whole of conception, thinking, and execution requiring highly developed sensorimotor and cognitive skills exercised reflectively and well. Yet each is also an evocative and understandable tract of human life and thought as experienced by somebody engaged with the world in an ongoing and dynamic way.

In each case, we are drawn away from just considering a mental exercise or calculus and into fluid rhythms of life in which something like 'calmness', 'pressure', 'style', or 'flair' are terms that naturally come to mind. Although in the operational cases of the general or doctor a more austere description may be aimed at, training plays a role here as do coaching and practice. Cricket is just one example of a sport where human over-thinking can have a negative effect. In each case, the agent is preparing another individual (the actor him/herself) for just such moments as that which is in the spotlight. In each case, a certain calm mental focus is important. Some of this conscious calm may be enabled by unconscious states such as the procedural memory that enables us to perform certain activities without having to think about the details of that performance.

Law-like causal accounts asserting natural kinds and biologically specified functionally related mental states and events are inadequate for an account of the human mind and the flexibility of adaptation revealed during our incarnate being in the world. This inadequacy prompted the rise of 'anomalous monism' (Davidson, 1982, pp. 207ff.). This was a naturalistic attempt to meet difficult conceptual problems in the philosophy of mind and language without accepting Wittgenstein's critique of earlier views of that philosophy as itself a prompt to do more than think again within what was a fairly widely accepted paradigm. The paradigm as it developed neither referred to, nor apparently recognised Hughlings Jackson's much earlier 'concomitance' or anti-reductive writing on psychology, psychiatry and neuroscience.

The difficulty with anomalous monism and its academic context, whether outlined by Davidson or any of those who followed him, was that it failed to give an adequate account of the anomaly concerned. The omission probably reflects the complexity and flexibility of the human neurocognitive system, as embodied in a living human being adapting to the rhythms of life as we live and share them within a human culture. Again, this embodiment and adaptation are located in an ethological and historical context, which may very well escape philosophical examination. In fact, Davidson later adopted a more inclusive position on the nature and content of mental events, similar in some respects to theories of the extended mind (Davidson, 2001; Rowlands, 2010). Yet he neglected the neural basis of mentality. The dynamic brain is therefore perhaps the source of the anomaly in a neurocognitive naturalism. Unfortunately, Freeman's work was not available to naturalistic thinkers of Wittgenstein's or the early Davidson's eras and therefore was not engaged with by them. We are not so limited and can begin to explore the *Dasein* of our flexible being in a way that builds on the dynamic neurophilosophy we have outlined.

Davidson's anomalous monism served as the conceptual start or opening strategy of a programme of philosophical responses to ambitious materialism and its reductive tendencies regarding mental life. But it leaves us wanting a more positive account of this life and its flexibility in a challenging world. That is what a neurodynamic account must offer if it is to satisfy us as being constructive for neurophilosophers wanting to go beyond the many unsettling questions left by disaffected sceptics of traditional conceptions of mind.

FLEXIBLE NEUROCOGNITION

It is time to begin to set out an account of flexible being that will respond to the familiar rhythms of human life in a way that does not involve simplification or notionally provoke an acerbic response from the fussier Wittgensteinians among us. Neurodynamics is up to that task as long as it takes account not only of the *philosophia perennis* that Ardley enjoins us to analyse. It also offers us an entree into both socio-cultural and ethical thinking. These form the core of the concept of *ethos* which partly bridges the conceptual gap between culture and ethics for a naturalist.

Much of this work will be done by pursuing Wittgenstein's later thought as it goes from logical and linguistic certainty to culture and value. If we accept that the mental is anomalous and therefore does not obey causal laws, then we are still devoid of a satisfying positive account of mental explanation and the psychological life of human beings (although ethology helps).

For example, we might invoke the case of a climber who feels himself about to fall during a climb (Gillett & Liu, 2016). He is roped to a woman who is not only his climbing companion but to whom he is engaged to be married. He may, let us say, be beset by doubts about his engagement and wish to be free of that particular promise. He realises he could save himself by cutting his *fiancée* free; but then she would tumble to her death. He knows that to be an honourable person, he must not do it but must face the mortal risk together with her. That thought so unnerves him that he loses his grip on the rope and she does fall. In this case, if we are looking for a mental explanation, we seem to have all the elements we need for a law-like or at least satisfying 'belief-and-desire' type explanation. He believed his fiancée would fall to her death if he let go of the rope, he desired to be free of both attachments, he so acted despite his hesitancy.

This type of account suitably quantified by some, is required for a functional complex of mental states causing an action. But such a cognitive account does not fully represent the actual situation. Psychological explanation is therefore not so easily rendered in belief-desire-propositional attitude terms. Such a psychology might be hoped for and championed by some philosophical theorists who like neat systems of explanation. Alas, we are flesh and blood and Aristotle, along with other classical philosophers, realised that *akrasia* (weakness of will) lurks within us all and may affect intentions at the most tragically poignant junctures.

The problem of working with an ontology of actions and events in mental life has been extensively discussed by one of us previously in relation to the event of a daughter coming home from school. That event has multiple coincident relationships to other events and aspects of situations or states of affairs. But there are no generalisable or remotely lawlike causal laws connecting those events (Gillett, 2008, pp. 36ff.). The intractability of any complete representation of many mental and other states and events in our lives and thus their intransigence in relation to systematic or scientific individuation and explanation leaves the philosophical stage open for an account of action of a more diverse type. When appreciated, that problem infects all philosophical discussion of mental or neurocognitive life in terms of specifiable states and events. That problematic realisation is at the core of neurodynamics with its living and inclusive relationship to the realities of human life in the world and the ways we have of going on within it.

The flexibility and fluidity of human neurocognitive life can be further illustrated by considering culture and creativity. Both of these areas of thinking take us far beyond the work of everyday survival or material success. They involve an expansion of the mind that touches on our human potential for the use of imagination and the development of techniques or skills which are not purely instrumental but have a broader significance. In fact, the techniques themselves may have an adaptive use and their extension to the limits of imagination or creativity (as in art) may refine them in ways that go beyond the everyday needs of the here and now. We see that, as we have mentioned, in blade-making and the idealised forms that appear when that skill becomes a prehistoric mode of Stone Age art.

A similar set of complexities and indeterminacies attend the identification of a neurocognitive event even as simple as seeing a mosquito land, for instance on a friend's arm at the beach. At that point, for those prone to over-thinking, complexity and the multiple possibilities that arise are salient and occupy much of our conscious space. That cognitive disability or predisposition means that a philosophically driven identification of actual neuroscientific states and their possible sequelae can be paralysing and prevent us from acting naturally and with a 'lightness of being' (as a *phronemos*). The ideas of a state of affairs and an intentional response in an active human life in the world is therefore simplistic and 'clunky'. It is also even more of a pipe dream than a realistic scientific entity if one subscribes to a workable causal theory of action (Gillett, 2008, 32ff). Propositional attitudes within such a theory are idealistic cognitive abstractions wearing pseudo-naturalistic 'causal drag'.

Every human culture has some expressive use of the techniques which in everyday life are adaptive within their ecological setting imbuing the life of the person concerned with a cultural twist (like 'a twist of lemon' in a drink or recipe). Thus, each culture has a way of working with nature to embody something characteristic of their own rhythms of life; such an ethos (among other things) is a set of ways of going on for ordinary purposes. This fact about human life, wherever it is found, suggests that we might usefully explore some themes arising in Wittgenstein's treatment of culture and value in his work of the same title (1998).

The very first remark *in Culture and Value* invokes complexity, personal history, familiarity with a situated experience, and the intricacies of developmental psychology.

> We tend to take the speech of Chinese for inarticulate gurgling. Someone who understands Chinese will recognise language in what he hears. Similarly, I often cannot discern the humanity in a man. (1998, #1914)

We have already referred to these intricacies in Chap. 3 by drawing on work by Vygotsky and Luria. Here ontogeny and the unique ethological situation of birth and personal history are included in the reading of personal development and weave together body and mind in a person's experience.

> It is difficult to tell a short-sighted man how to get somewhere. Because she cannot say to him: "look at that church tower ten miles away and go in that direction." (1998, 1e)

> No one can think a thought for me in the way no one can don my hat for me. (1998, 2e)

Note that this is 'thinking a thought' with its idiosyncratic resonances. It is distinct from the way we have discussed it in relation to philosophical logic and metaphysics. It is also an extremely astute observation about a human life slightly disabled in a way that borders the normal (so statistically prevalent that it may seem elitist to regard it as a disability). The requirements of systematicity and unambiguity rule in metaphysics and philosophical logic as Frege and the Wittgenstein of the *Tractatus* had argued. The later Wittgenstein notes that these are a 'stripped down' core and not what we commonly refer to as 'thought' in all its psychological richness.

> You cannot lead people to what is good: you can only lead them to some place or other. The good is outside the space of facts. (1998, 3e)

These few remarks are close to the beginning of *Culture and Value*, and yet one can already see the fluidity and flexibility that underpin the

complex strands of thought involved. One particularly notes the complexities of trying to treat 'the good' as something definitively identifiable
in descriptive terms or by specifying an objective cognitive location.

It is worthwhile and instructive to construct a syllogism encapsulating
Wittgenstein's point while acknowledging that syllogisms are abstractions
which artificially formalise the cognitive process. Wittgenstein alluded to
this in remarking about our varied and subtly nuanced ways of appreciating a thought. Notice that his hesitation remains even when we have
understood 'a thought' in a way sufficiently like others to allow logic and
philosophical reasoning to proceed.

1. Complexity and the dynamic holism of neural life at any moment
   precludes talk of events.

Therefore

2. Along with talk of events we lose event-causal neurocognitive
   explanation.
3. Informed contemporary neuroscience precludes an event-causal
   reductive metaphysics of mental life.

Conceiving of a neurodynamics of the rhythms of life is as complex and
elusive as many other natural phenomena. When we try to individuate
them in terms of specified complexes of identifiable components apt to be
captured by our post-industrial causal laws, they slip from our grasp in the
same way that frustrated the *Tractatus'* scheme and its austere ontology.
That ontology was an idealisation of a logic-driven philosophy. But life is
'fluid' and seamless, like currents in a river. This metaphysical difficulty
should not stand in our way as we try to formulate a realistic naturalism
about human neurocognition as the basis of our flexible and adaptive
behaviour. These problems beset naïve realism about quantum physics *a
fortiori.*

Genetics, Epigenetics, and Gender Identity

We can gather the shaping influences on a human being broadly under
two groupings: genetic and epigenetic. The latter are often scientifically
conceived and studied, particularly in biology and medicine, as dealing
primarily with intracellular mechanisms and chemicals that influence the

expression of genes. We shall broaden the scope of that discussion for the present purpose to return to the base meaning and include contextual and other environmental factors. These factors affect the way genes are modified in their action to shape the living characteristics of the human organism. As we broaden the discussion, we should note that human beings themselves, considered as autopoietic beings, sometimes have a role here (think for instance of the choice of marital partners and the consequent genetics of children). Once this broader scope is taken into the account, the way will be opened to properly consider the many ways that discursive and other influences affect the historical development of the human race.

Throughout human history, and presumably prehistory before that, human beings have altered their context of life so that their biological limitations could be overcome. We do not have the sharpest teeth, nor the most effective claws, nor the fastest natural means of locomotion, nor the most effective insulation for our bodies. Yet we have overcome those limitations. We have developed ways of doing things and the technology to supplement our natural abilities. We make sharp blades even from stone. We have ways of preparing and eating a huge variety of food sources that diminish their possible hazards, running down our prey or otherwise overcoming their advantages in terms of speed of travel, and making even sub-Antarctic conditions habitable. And all that is not to mention means of crossing oceans and exploiting what is to be found in their depths. We have means of using the air to transport us over vast distances whatever the geographical difficulties to be overcome, and even ways of making devices that allow us to become deadly predators able to strike our victims without exposing ourselves to any dangers that they pose.

There is a curious interaction between these, partly artefactual and autopoietic, factors and human evolution. Certain genetic predispositions embodied in physical types provide some individuals with advantages over other human beings in particular contexts. For instance, the ability to sustain concentration over an extended period of time despite distractions is probably genetically determined to an extent, with a defect in that ability being something like impulsivity or even attention deficit hyperactivity disorder (ADHD) and related conditions (Faraone & Larson, 2018). We recognise one of these as a pathology, but the others merely seem to be an individual variation and to require a certain context of life, employment, or culture-related success within a human society to be useful.

Something like this fact is evinced by the common quip that Oxford and other centres that are home to distinguished academics contain the highest concentrations of autistic people in the world (Frith & Hill, 2003). One can recognise the irony and, with a modicum of insight, perhaps come to realise that this saying is not only apt but, to some extent, applies to oneself as an academic author. The trick is to moderate the autism to a sufficient extent that one's work becomes accessible beyond the peculiar individuals on whom it draws (Vorstman et al., 2017). It is a further step away from that curious distinction to be genuinely publicly accessible. There is also the 'small' problem of living an ordinary human life with all its everyday concerns. In this, sensibility to the needs and limitations of others, and perhaps insight about one's own, is a blessing which Wittgenstein seemed to show to a fault. Such insight is not the lot of all academics. But it is the lot of some, one of whom said to one of the authors that he found casual chats difficult at conferences because "My mind works so fast other people have difficulty keeping up with me".

Wittgenstein again:

> I ought to be no more than a mirror, in which my reader can see his own thinking with all its deformities so that, helped in this way, he can put it right. (1998, 18e)

And again:

> I don't believe I have ever *invented* a line of thinking. I have always taken one over from someone else. I have simply straightaway seized on it with enthusiasm for my work of clarification. ... What I invent are new *similes*. (1998, 19e)

Here we have a rare insight into the flexibility and creativity that Wittgenstein notices in academic life—his ethological niche. Parallel remarks could be made about carpentry, cooking, the fine arts, even hunting. Each has its own mix of adaptation of the existing human resources and creativity of new ways of going on based on the holistic flexibility of the human neural net and its interaction with the environment. The interwoven nature of biology and human intervention, invention, and artifice under socio-political pressures is poignantly illustrated by issues associated with gender in modern society and human life in general.

Most sports codes have clear differences for male and female world records associated with the particular sport-related performance involved. These differences are almost certainly based on biology. In many other areas of life, though, it is much more difficult to say what is due to differential biological aptitude and what is due to cultural expectations and what is developed consequent on gendered expectations, roles, and modes of upbringing.

It is tempting to think of sex as binary—male and female—based on reproductive role. By and large, that seems to be true throughout the animal kingdom until we come to human beings. We now recognise gender diversity as an increasingly evident reality throughout the world, and therefore the old binary categories have been superseded. Despite cultural resistance in certain contexts and a wide range of religious positions, gender diversity seems widespread among both biological males and biological females. There is little question that it is the result of biological factors influenced to some degree by socio-cultural factors. But it is also due to individual factors of uncertain nature and origin. These considerations need to frame our further discussion of the discursive life of human beings of which the soul or individual *psyche* is definitely a feature.

It is worth summarising the argument in terms of a further syllogistic construction.

> If gender was determined by biological sex, then it would be binary.
> Gender is not binary.

Therefore:

> Gender is not determined by biological sex.

The syllogism is, of course, an austere abstraction from the complex reality of sex and gender. Like all human phenomena, it is a history deeply entwined in tradition, culture, and religion. That history is complex, and any summary must be both selective and tendentious even if it is motivated by a desire to be neither judgemental nor inflected by over-simplification or a socio-political agenda. The task is difficult for any philosophical inquiry.

Just such a balancing exercise requiring judgement and careful argument and evaluation understandably leads some to take refuge in syllogisms. But as we have seen, syllogisms are not purely logical but themselves deeply affected by the choice of premises and the resonances of the language used to enunciate them. Therefore, the history must, to invoke a telling contemporary innovation, being neither a 'history' nor a 'her-story' but something else, perhaps indefinable.

Bonnie Smith remarks: "The rise of feminist movements across the globe also motivated the writing in women's history over the last 250 years, even as the narrative of that history felt the influence of national liberation, civil rights and postcolonial activism". (Smith, 2013, 2000). Her work makes the multifaceted complexity of discussing the issue clear but merely scratches the surface of the fraught human agendas that interweave within this discussion. A recent discussion of gender diversity surveys the several ways in which gender diversity can influence scientific discussion and discovery (Nielson et al., 2018). The authors mention that gender diversity might apply to research teams, research methods, or research questions and that each of these may introduce voices to the discussion which need to be heard but may also introduce conflicts and problems. Not the least of those problems concerns the way in which scientific merit is measured and the extent to which 'scientific' methodologies and the assessment of findings are influenced by a post-industrial, predominantly causal, conception of discovery, and significant natural explanations. There is no doubt that this paradigm is now challenged. But a unified philosophy of science is difficult to formulate, and many fear that a type of anarchy and disorganisation in the service of conflicting socio-political programs threatens scholarship. The authors also mention that multiple intersecting influences and socio-historical forces can shape the agenda and influence discussion in this area.

The thought that 'it is difficult to keep your head when all around you are losing theirs' is only rendered more daunting when one also recalls that quip: 'there is nobody sane around here except you and me, and sometimes I wonder about you'. The latter remark should enjoin a modesty about philosophical reasoning. Nevertheless, we have no choice except to discursively illuminate any subject of enquiry as a retreat to syllogisms in an admission of defeat.

Wittgenstein himself seems to have taken inspiration for some of his most profound ideas from diverse sources: "One day when Wittgenstein was passing a field where a football game was in progress the thought first stuck him that we play Games with words. A central idea of his philosophy, the notion of a 'language game', apparently had its genesis in this incident" (Malcolm, 1965, p. 65). One is immediately struck by the aptitude of the thought. A game can be serious, and the rules strictly enforced or more casual, with only a crucial skeleton of them attended to. Language is similar. The present work, like most philosophy, is more on the 'strict' side of thought, and something like a poem or a 'rap' otherwise so that other rules and values come into play as fewer rigid manifestations mark it. One might even say: 'The rhythm is all'.

CONCLUSION

We have discussed neurocognition in terms of neural rhythms that are embodied in individuals whose thought is shaped by different aspects of culture. This is an elaboration of Wittgenstein's examination of culture, of value, and of meaning in its broad sense, in his later work. We have argued that the phenomenology and content of consciousness cannot be explained, or explained away, by reductive materialism because what it is like to perceive an object or experience an event, as well as the actions we intend to perform, and how we act and interact with others, cannot be expressed in materialist terms. Promising has afforded an instance of a human practice that, with its cognitive, enactive, and socio-cultural complexities distances us from other animals with whom we share many attributes and the neurocognitive basis in which they are embedded and enabled.

Indigenous perspectives, like the Māori perspective we have noted, provide a richer texture to the fabric of flexible being-in-the-world by showing the impact of the immediate environment and culture on thought and other forms of mental life. Indigenous cultures also have comprehensible and nuanced 'lore' which many Europeans have been drawn to and thereby complexified their being and belonging. This cultural texture also includes the complex relations between genetics, epigenetics, and gender identity and its role in sport and other dimensions of our lives. But we must be cautious and extend an Aristotelian mean of caution into the chapters to follow, enforcing an attention to scientific norms while using Frege's 'pinch of salt' to avoid extremes and dogmatism. With that caution in mind, we can turn once more to a philosophical discussion of discourse and its complexities for the discursive mind (Harre & Gillett, 1994). This is the focus of Chap. 5.

REFERENCES

Abbar, M., Demattei, C., El-Hage, W., et al. (2022). Ketamine for the acute treatment of severe suicidal ideation: Double blind, randomised, placebo-controlled trial. *BMJ, 376*, e067194. https://doi.org/10.1136/bmj.2021.067194

Anscombe, G. E. M. (1978). Rules, rights, and promises. *Midwest Studies in Philosophy, 3*, 318–323.

Ardley, G. (1968). *Berkeley's renovation of philosophy*. Martinus Nijhoff.

Bayne, T. (2009). Perception and the reach of phenomenal content. *The Philosophical Quarterly, 59*, 385–404.

Benedetti, F. (2014). *Placebo effects* (2nd ed.). Oxford University Press.

Bolton, D., & Gillett, G. (2019). *The biopsychosocial model of health and disease: New philosophical and scientific developments*. Palgrave Macmillan.

Carhart-Harris, R., Roseman, L., Bolstridge, M., et al. (2017). Psilocybin for treatment-resistant depression. fMRI-measured brain mechanisms. Scientific Reports, 7, 13187. https://doi.org/10.1035/srep41598-017-13282-7

Davidson, D. (1982). *Essays on actions and events*. Clarendon Press.

Davidson, D. (2001). *Subjective, intersubjective, objective: Philosophical essays, Volume 3*. Clarendon Press.

De Haan, S., Rietveld, E., & Denys, D. (2015). Being free by losing control: What obsessive-compulsive disorder can tell us about free will. In W. Glannon (Ed.), *Free will and the brain: Neuroscientific, philosophical, and legal perspectives* (pp. 83–102). Cambridge University Press.

Dennett, D. (1987). *The intentional stance*. MIT Press.

Faraone, S., & Larson, H. (2018). Genetics of attention deficit hyperactivity disorder. *Molecular Psychiatry, 24*, 562–575.

Frith, C. (2008). The social functions of consciousness. In L. Weiskrantz & M. Davies (Eds.), *Frontiers of consciousness* (pp. 225–244). Clarendon Press.

Frith, U., & Hill, E. (Eds.). (2003). *Autism: Mind and brain*. Oxford University Press.

Fuchs, T. (2011). The psychopathology of hyperreflexivity. *The Journal of Speculative Philosophy, 24*, 239–255.

Gillett, G. (1985). *Persons and their mental functions*. Unpublished Oxford D. Phil thesis.

Gillett, G. (1988). Consciousness and brain function. *Philosophical Psychology, 1*, 325–339.

Gillett, G. (1992). *Representation, meaning and thought*. Clarendon Press.

Gillett, G. (2008). *Subjectivity and being somebody: Human identity and neuroethics*. Imprint Academic.

Gillett, G. (2009). *The mind and its discontents* (2nd ed.). Oxford University Press.

Gillett, G. (2016a). Culture, the crack'd mirror, and the neuroethics of disease. *Cambridge Quarterly of Healthcare Ethics, 25*, 634–646.

Gillett, G. (2016b). Neural plasticity and the malleability of pain. In S. van Rysewyck (Ed.), *Meanings of pain* (pp. 37–54). https://doi.org/10.1007/978-3-319-49022-9_3

Gillett, G. (2018). Concussion in sport: The unheeded evidence. *Cambridge Quarterly of Healthcare Ethics, 27*, 710–716.

Gillett, G., & Liu, S. (2016). The four levels of free will. *Journal of Cognitive Science, 17*, 167–198.

Glannon, W. (Ed.). (2015). *Free will and the brain: Neuroscientific, philosophical, and legal perspectives*. Cambridge University Press.

Harre, R., & Gillett, G. (1994). *The discursive mind*. Sage Publications.

Higgins, D., Baker, D., Vasterling, J., & Martin, A. (2018). The relationship between chronic pain and neurocognitive function: A systematic review. *Clinical Journal of Pain, 34*, 262–275.

Hughlings Jackson, J. (1887). Remarks on the evolution and dissolution of the nervous system. *British Journal of Psychiatry, 33*, 25–48.

Jacyna, C. (2011). Process and progress: John Hughlings Jackson's philosophy of science. *Brain, 134*, 3121–3126.

Kant, I. (1790/2007). *Critique of judgement* (J. C. Meredith, Trans.; N. Walker, Ed.; Rev. ed.). Oxford University Press.

Knox, T., Anderson, L., & Heather, A. (2019). Transwomen in elite sport: Scientific and ethical considerations. *Journal of Medical Ethics, 45*, 395–403.

Kripke, S. (1980). *Naming and necessity*. Harvard University Press.

Malcolm, N. (1965). *Ludwig Wittgenstein: A memoir*. Oxford University Press.

Marsden, M. (2003). *The woven universe: Selected writings of Rev. Māori Marsden* (T. A. C. Royal, Ed.). Estate of Māori Marsden.

Mashour, G., Roelfsema, P., Changeux, J.-P., & Dehaene, S. (2020). Conscious processing and the global neural workspace hypothesis. *Neuron, 105*, 776–798.

Mikkola, M. (2009). Gender concepts and intuitions. *Canadian Journal of Philosophy, 39*, 559–584.

Nagel, T. (1974). What is it like to be a bat? *Philosophical Review, 98*, 435–450.

Nagel, T. (1986). *The view from nowhere*. Oxford University Press.

Nielson, M., Bloch, C., & Schiebinger, L. (2018). Making gender diversity work for scientific discovery and innovation. *Nature Human Behaviour, 2*, 726–734.

Rowlands, M. (2010). *The new science of the mind: From extended mind to embodied phenomenology*. MIT Press.

Sartre, J-P. (1958). *Being and nothingness* (H. Barnes, Trans.). Methuen.

Scanlon, T. M. (1990). Promises and practices. *Philosophy & Public Affairs, 19*, 199–206.

Schiller, F. (1795/2016). *On the aesthetic education of man* (A. Schmidt & K. Tribe, Trans.). Penguin.

Smith, B. (2000). *The gender of history: Men, women, and historical practice*. Harvard University Press.

Smith, B. (2013). Gender I: From women's history to gender history. In N. Partner & S. Foot (Eds.), *The Sage handbook of historical theory* (pp. 266–281). Sage.

Smith, W., & Sisti, D. (2021). Ethics and ego dissolution: The case of psilocybin. *Journal of Medical Ethics, 47*, 807–814.

Snowdon, P. (2008). Strawson on philosophy—Three episodes. *South African Journal of Philosophy, 27*, 167–178.

Strawson, P. F. (1974). *Freedom and resentment and other essays*. Methuen.

Van Rysewyck, I. (Ed.). (2016). *The meanings of pain*. Springer.

Vorstman, J., Parr, J., Moreno-De-Luca, D., et al. (2017). Autism genetics: Opportunities and challenges for clinical translation. *Nature Reviews Genetics, 18*, 362–376.

Wittgenstein, L. (1921/1974). *Tractatus Logico-Philosophicus* (D. Pears & B. McGuinness, Trans.). Routledge.

Wittgenstein, L. (1998). *Culture and value* (G. H. von Wright, Ed.). Revised edition. Blackwell.

Zeman, A. (2008). Does consciousness spring from the brain? Dilemmas of awareness in practice and in theory. In L. Weiskrantz & M. Davies (Eds.), *Frontiers of consciousness* (pp. 289–322). Clarendon Press.

CHAPTER 5

Being Discursive

We can begin a reflection on being discursive by picking up from the point on which we closed our reflections on language and games. We should consider yet again why we should resist certain kinds of systematisation and post-colonial pseudo-scientific theorising about language. We have noted that the complexity and fluidity of language embroils any study of the mind and the philosophy of thought in our diverse ways of going on as beings in the world related to each other in many intertwined and socially situated ways. As we consider 'discourse' and the way it is interwoven with the rhythms of human life in the world, so we are reminded again of J.L. Austin's *How to Do Things with Words* (1962).

In contrast to many theories of language based on 'denotation' and 'predication' and increasingly complex formal elaborations of that basic structure (Lycan, 2019; Stalmaszcyzyk, 2021), Austin sought to return language to its practical engagement in everyday life. There it is used to do things rather than merely denote or describing items and their properties, where these items include denumerable actions and events. We make promises, we get married, we enunciate and communicate laws of scientific, civil, and criminal types, we describe and evaluate works of art, coach football teams, and so on. Thus, language is a form of action that is woven into our everyday ways of going on variously rule-bound and/or playful, depending on the occasion. We are therefore enmeshed in the complexities of which Wittgenstein astutely reminds us in relation to language.

G. Gillett, W. Glannon, *The Neurodynamic Soul*, New Directions in Philosophy and Cognitive Science, https://doi.org/10.1007/978-3-031-44951-2_5

Austin himself provoked a whole philosophical industry based on the idea of 'speech acts' (Searle, 1969; Fogal et al., 2018). But his overall point is that language is an active part of our complex and interactive being-in-the-world. It enables certain aspects of that being and facilitates others. It is far more important as a reflection on life and cognition than any historical 'moment' in analytic philosophy.

Discourse, Persons, and the Unity of Consciousness

It will be useful to revisit a topic in neurophilosophy which to date has not really had an organic place in the discussion: The topic of brain bisection on which one of us has written extensively in the past (Gillett, 1986). In its heyday this topic inclined some to believe that the unity of the human mind was a secondary phenomenon constructed on the basis of a *de facto* and *de dicto* (or vice versa) integration in terms of biological coexistence within an organism of essentially discrete neurocognitive functions. These were actually articulated through causal connections which could be studied and theorised as functional components of a complex neural 'machinery'. The unity of the self was a secondary and, to some, an illusory construct. Real complexity based on entangled life in the world and the associated philosophical theory was not part of that picture. We therefore have a puzzle to be addressed by argument and a challenge to be confronted by clinical experience.

The challenge:

1. Human beings seem to be unitary subjects.
2. Physical changes, when confronted by careful 'experiments' (some 'natural' such as strokes or surgery), have divided mental functions.
3. (Therefore) unity of the mind is an illusion and not a reality.

Except, one might observe:

Human beings can recover practical unity of function if returned to everyday settings and learn to recover their self-integration.
Therefore, one might rather conclude:
3a. Human beings are naturally integrated subjects adapted variously and individually to the world. But that adaptation can be compromised by pathology, as can any neurocognitive function.
HA corollary might be: 4. Disruptive interventions can impair bottom-up or top-down integration.

However, that broader integrative and involved-in life view did not fit with the prevailing self-images of the age with their reductive prejudices (MacIntyre, 1989).

The present inclusive and holistic view seems to be important in the history of neuroscience. Some of its great pioneers thought in terms of an understanding of the human neurocognitive system as having complex and integrated rhythms of function. As Hughlings Jackson, Luria, and others have pointed out, these rhythms are adapted variously and holistically to everyday life in the real world. Some of the defects that appeared in the split-brain patients in a neuropsychological laboratory by manipulation of information available through the visual system were not evident in the everyday life of the individuals tested. Although this turned out to be a severe blow to bundle theories of mental states and events in various forms and also to some forms of post-industrial functional localisation, it was breezily waved aside by 'language on holiday' in its neuro-philosophical form (*Philosophical Investigations*, 51). Again, Wittgenstein was prescient about the misconceptions of scientific psychology, though not, of course, about these experiments and their interpretation (anachronism does not become us, however appealing to a tendentious mind).

We have inherited these post-industrial scientific views. These have even penetrated the discussion of the neurophilosophy of quantum physics (Schwartz et al., 2005; Tarlaci & Pregnolato, 2016). Here 'mathematicism' about cognition and functionalism in neurophilosophy have spawned theories such as QBism that have emerged in the fraught discussion that ensued when scientific realism met quantum theory. The assonance between QBism ('Quantum-Bayesianism') and cubism in art is clever because the art movement also abstracts from realistic images by substituting abstract forms which render 'the natural' in distinctly human and artificial forms.

The everyday dealings of a human being with life in the real world in which they encounter objects can be tracked in space and time. We track them using our holistic sensory motor system as the basis of neurocognition. This is consistent with to the developed analytic philosophy of thought and language after the *Tractatus*. In thinking, as we have conceived it, a human being develops an integrated neurocognitive lifeworld as part of that complex and flexible orientation to the diverse demands of that world. Thus, the split-brain experiments which seemed so startling in their implications in their day could be relegated to curiosities in the philosophy of neuroscience in terms of a more inclusive philosophical understanding of real embodied human neurocognitive life.

The enthusiasm that attends the voices of any age is apt to produce in philosophers an interest that has generated intense and highly intriguing debates about abstract topics such as personal identity and the unity of consciousness (Eccles, 1970; Nagel, 1971; Shaffer, 1977; Parfit, 1984, pp. 378–9). The impaired but integrated struggles of a person dealing with a damaged brain were not included in these debates and not well understood in philosophy. We have seen them discussed in medicine and neuropsychology (e.g., Luria, 1987). We would not be the successful species that we are unless we had learned to attend to and selectively integrate the many cues and clues that our situated historical life affords cognition. What is more, none of us is alone in that learning because we communicate discursively about objects, their properties, and the events in which they are encountered with others who are not impaired. Impaired and unimpaired minds are both necessary to understand the neuroscience and psychology of agency and adaptability.

Being discursive therefore guards us against the many dangers to which fragile, incarnate, human life exposes us in terms of our relatively quick reactions, our vulnerability, and the associated morbidity and mortality. It is a testament to the integrating and restorative abilities of human beings that such pitfalls can be negotiated and recovered from, even when they seem impossible to survive as an intact and living player in the game of life.

Social Integration and Spirit: Recovery from a Brain Injury

Consider the case of Nick Chisholm, a rugby player in Dunedin, New Zealand, who suffered a brainstem stroke during a game. This developed into a form of locked-in syndrome (LIS), where one is conscious but almost completely paralyzed. Eyelid movements are often the only remaining voluntary motor control. Like Luria's patient Zazetsky more than 60 years earlier, Nick's recovery of some motor functions and the strength of will he put into it are a remarkable testimony which has inspired many to recover from that dire neurological condition (Chisholm & Gillett, 2005). Correspondence from the patient's relatives and doctors from all over the world was received by the authors. Readers were encouraged by the fact that a condition many failed to diagnose and caused the patient to be regarded as hopelessly brain-damaged could lead to some degree of recovery.

This was published in a major medical journal. In that first-person account by a person who had lived through an "horrific experience that he would not wish on his worst enemy" we see a character, replete with friendships and remarkable spirit, who overcame one of the most terrible afflictions that any human being could face. So terrible is that situation that many people in various parts of the world find themselves unable to face the future and take the mortal decision to end their lives or have the decision effectively taken for them by well-meaning and compassionate healthcare professionals and relatives (Halan et al., 2021).

The authors of that survey review the global statistics of LIS and the common outcomes which make it clear that Nick Chisholm was an exceptional case. He is also, as noted, a testimony to not only his own human spirit but also that of his loved ones, friends, and caregivers. *Autopoiesis* in his case, as in all human beings, was a profoundly socio-political and intentional affair requiring both a cultural setting of valuing and supporting human individuals regardless of wealth or status. It showed 'wrap-around' community values. For instance, his regional rugby association gave him life-long admission to games. Nick's case reveals the real-world face of neurophilosophy in general and neuroethics in particular. It shows how a person's resilience in facing and overcoming the challenges of a severe brain injury can promote flexibility and adaptability to drastically changed circumstances. Resilience occurs in context. It depends not only on the individual's own psychological resources, such as the will in trying to recover from adversity, but also on the social and cultural community that supports and helps to realize this recovery (Southwick et al., 2023).

Certain bodily images become entangled with us at the core of our being and are essential to our experience of embodiment. The phenomenology of embodiment can be salutary in health but, alas, deleterious in disease. It is where the 'fly' cannot be clearly distinguished from the 'fly bottle'. One can imagine a particularly reflective fly wondering just what the strange creature is that he sees as he tries to look at the world outside the bottle. Yet he seems insubstantial as he tries to initiate movement. Imagine that the fly arrived at a self-conception, but over time the task of integrating with the world became more challenging. Such a confused image sometimes confronts a patient who has suffered a life-changing self-image adjustment from a cancer diagnosis, brain injury, or other biomedical event. The revision and adjustment of the self can be paralytic, and the strength to carry on may seem out of reach for the affected person. They may not have the strength of character and will needed to do this on their own. The paralysis can be more severe if disease or injury causes them to lose their discursive capacity and ability to communicate their experience to others.

Integrated and flexible being-in-the world, which includes caring interpersonal relationships and the discursive supplementation of life skills by others who become involved in each other's lives, makes recovery from a devastating neurological condition a real possibility. This is something that even a person in a locked-in syndrome can attain in a supportive and equipped context. The result is an inspiration to all of us and an appreciation of the way that discursive being serves as an entree and foundation for a full human life. In light of this, we can consider a further 'tweak' on the relevant syllogistic reasoning, building on that previously enunciated.

1. Brain damage can cause the dissolution or disruption of human neurocognitive function.
2. A human being is a bundle or complex of such functions.
3. A damaged human being can reintegrate him or herself as a unified being in the world.

(3) implies that a person is more than just a collection or bundle of neurocognitive states and functions. Therefore (2) must be interpreted carefully or else is false. The many thought experiments involving split brains and teletransportation used by Derek Parfit and others to undermine the conception of a unified human neurocognitive life engaged with an objective context are therefore often misleading. They fail to appreciate or even ignore all the integrated features in a fully articulated neurophilosophy.

Wittgenstein warns of "when language *goes on holiday*" (Wittgenstein, 1953, #38). Such thought experiments are an example; "Such fun" one might say. But philosophy for him was not so; it had work to do. Therefore, these are fantasy and fun, not philosophy. The job of philosophy is to reaffirm the complexity and objectivity of our relatively unified life in an objective world of discursive subjects living among others and to better equip our minds for that job. The problem of abstracting from lived experience in these experiments is also found in the philosophy of language. In the following relevant passage, Wittgenstein is critiquing an abstract idea of 'denotation' separated from ordinary everyday discourse as in formal 'structural' theories of language or meaning and intending as 'inner events'.

And, also, the word "this" and a name often occupy the same position in a sentence. But it is precisely characteristic of a name that it is defined by means of the demonstrative expression "That is N" (or that is called 'N'). But do we give the definition: "That' is called 'this'" or "This is called 'this'"? (1953, #580)

He goes on explicitly to critique the very idea of an 'occult process' bringing into doubt that philosophical brain-child, the oft-posited concept of mental events.

In this world, we identify and re-identify things over time and experience subject to realistic constraints which apply to organisms of our kind. We note their continuity as historical and extended phenomena in our ethosphere. We develop terms to assist each other in this cognitive work of identification and characterisation. We weave around some of those terms a scientific understanding which normatively constrains their use so that our collective thought does not come adrift from situated and observed reality at behest of 'language on holiday'. Real language as used when not 'on holiday' remains firmly grounded in that reality in which we live and move and have our being.

THOUGHT EXPERIMENTS AND THE NATURAL WORLD (REVISITED)

This naturalistic realism has caused some philosophers, such as Katheen Wilkes (1993), to be deeply suspicious of thought experiments because they are artificially detached from the natural world. We have noted that the 'experiments' deal with abstract conceptions unconstrained by complex natural contingencies. We honour the corrective historical voices when we keep our thought experiments real, that is, grounded in the world of our lived experience.

Consider a common imagined situation which is culturally plausible, especially in the urbanised world where, increasingly, we do not see nature as it actually exists around us. An urban child, call him Charlie, might be given a number of tadpoles but not really understand what they are. He goes on holiday leaving them in a tank for a neighbour's daughter to 'feed his little fish'. When he returns, they have become frogs, as tadpoles naturally do. Charlie does not know that in nature these 'little fish' undergo metamorphosis into frogs, and he thinks that the creatures now in this garden pond have eaten the 'little fish' to which he was becoming emotionally attached. Absent an acquaintance with the intricate and inter-woven complexities of metamorphic development, his mistake is quite understandable. Of course, discourse means that he can learn 'the truth'. One could say that truth is a cognitive victim of an urban lifestyle for children like Charlie, unless we do our part as educators. As such, many well-trained primary school teachers make the gathering of tadpoles for an educational purpose part of a child's experience. That experience, among others, re-engages the pupils with a common wonder of the natural world. This has the socio-political effect of attuning the child to nature and its regularities, alerting them, among other things, to the devastating effect of climate change. Who knows, perhaps one day all the tadpoles will be gone.

Science discloses the mechanisms and complexity underlying amphibian lifeforms and their metamorphoses. But ordinary everyday life assures us that it is part of nature, whatever we think. Science builds on such contingent human contact with the world and often does so in ways that show the world contains many things, the complexity of which we can hardly comprehend. This fluid complexity is what underpins our discourse and its logical or other availability to human thought. It spawns theories which attempt to render it in terms we can reason about. But that accessibility to reason, and with it the interwoven higher cognitive attributes of the individual, is itself very difficult to codify. For instance, a man might try to clean the kitchen of his house to the high standards that his partner usually attains but, as often as not, fall short. There could be a number of reasons for this task-related deficiency. He could be lazy or resentful of the fact that she so often fails to reward him. Or, the partner might observe in giving limited commendation to his attempts, "You do not have the eye". She might qualify that mild reproof by saying "That takes years of experience; you can't learn it overnight". This remark itself points to this skilled observation that develops through a multidimensional life in the world replete with skills, particular norms to uphold, and aims to be achieved (albeit distorted by human failings).

Some examples of culturally developed and attuned abilities combining genetic and epigenetic influences are:

1. The recognition of good 'black American' jazz music ("What makes good jazz?" Man, you gotta ask, you ain't never gonna know!"
2. Knowing quality flamenco dancing (or *capoeira*).
3. The complex relationship between *mana* and *aroha* in Māori culture.

The entanglement between human being, history, and context is how discourse arises and develops. We become discursive beings who sometimes trap ourselves in puzzles built from our partial knowledge of these complex entanglements. We opt for biological or social abstractions and create scientific generalisations, such as genetic explanations, to simplify a scientific account of ourselves. We distort complex metaphysical conceptualisations in this way and pose questions which can only be answered by accepting some device such as statistical levels of significance against which scientific beliefs can be measured.

The need for such abstractions exploits the dichotomy between what is actually possible and what our limited knowledge makes seem possible. In a world where 'language is on holiday' from its normal task of enabling us to adapt to reality as it encloses us, wondrous things can seem to happen. The interwoven complexity and dynamism of life in the world and the science to which it gives rise has led one notable indigenous thinker to coin the phrase "the woven universe" (Marsden, 2003), linking our thought in all its complexity to weaving. This time-honoured way of fashioning an important part of our human life immediately makes the point that scientific thought is a human creation but neither fanciful nor idealistic. Rather, it is interwoven with human techniques in general.

The practical cognition of indigenous people does not achieve significant recognition in modern scholarship. The latter often proceeds on the basis of the kind of abstraction to which analytic philosophers are prone. The resulting intuitive naturalism of indigenous thought, which may seem to some to be 'primitive', can be seen as remarkably sophisticated when it comes to such tasks as surviving in the wild—the Australian outback for instance—or living in a dynamic harmony with nature.

Such a thought world reserves counterfactual and magical thought for matters of religion and mystery of 'deep' importance. Such cognitive modesty is of a type that preserves the moral and evocative aspects of human neurocognitive abilities and invites a different kind of understanding of their content. We are complex embodied creatures whose lives are interwoven in complex and endlessly creative and flexible patterns which engage us with others and the world in an ongoing and dynamic way. Unlike a program or device that seeks to emulate the human mind, all sorts of things can be part of what emerges from the real embodied and embedded mix. We play football, drop bricks on our feet, and engage in artistic creativity and evocative discourse. These things to some degree or another are a vital part of various aspects of us in varying ways and must create a place for themselves in the human neural nets of those concerned.

A further illustration of human neurocognitive complexity can be found if we get real about so-called 'qualia'. Think about the human grasp of a concept like 'pink'. This seems to be based on a simple visual connection between an individual and the world and is driven by spectral features of visual experience. But if we think again, we realise that 'pink' has a complex role in human development and cultural experience as a whole. A child learning what 'pink' is must learn in discourse with adults to apply the term and master the concept by attending to the hue of a visual presentation, rather than its shape, colour, context, or any other sensory feature present at the time. Only so will he or she come to grasp what 'pink' refers to, far less its many cultural associations. These are complex and interwoven, so that at a certain point in Western cultural history it was not regarded as completely 'kosher' to dress little boys in pink and one is subtly encouraged to drift towards other colours such as blue or brown in order to teach them what it is to be typically 'male'. One must be careful of this kind of reflection on 'pink' in the current social climate where gender diversity is accepted but is also a heavily cultural judgement.

Thus, our cognitive mastery, even of simple sensory semantics and apparently related to individual visual experience, is complex and ramified throughout the human engagement with the world of others. The woven universe is a far better route to an understanding of human discourse than any supposed contrast between ideas and reality. Whether analysed in terms of an individual or a group, 'ideas' and 'reality' are still notionally and logically separable from our dealings with things together in the world. We learn to do things with words in ramifying and intersecting ways that cognitively enable us to master human techniques which enhance our adaptations to real contexts.

The woven universe entangles and enables us with complexities and patterns of being in the world wherever one looks. The patterns interact and support one another creating a whole which is difficult to untangle. Unfortunately, the woven universe, in terms of the selective blindness of some weaves, can so disrespect the real world and its contingencies that it threatens human life itself through arrogance and over-consumption. The actual woven universe in which we have our being in one that we share with our many families of belonging, whether relatively close or relatively distant. Relatively close members are such beings as 'brother wolf' and 'sister deer' whom we must respect even if, at times, we end their lives.

More remote are such creatures as trout, or eels, or even going further in terms of distant lineages, massive pine trees, homely apple trees, and the berries that help to sustain our lives. Even more remote in an evolutionary sense, while still sharing our most intimate spaces of being, are bacteria, fungi, and microorganisms with which we co-inhabit the world and which, in some cases, inhabit members of the close family.

DISCOURSE AND COMMUNITY

Here we need to make a point about discourse, objectivity, and community and the complex triadic relationship between them. This can be obscured in localised philosophical debates such as that between individualism and communitarianism. An example suffices to clarify this relationship. One of us lives in New Zealand and has a property which features a wild garden. In that back garden, there is a malignant creeper which tends to kill other plants by growing on them and then sucking and strangling the life out of them. My wife and I call it 'evil neevil', though that name is not its botanical name nor part of any properly public discourse. Nevertheless, the name serves to refer to the pest and even conveys a 'sense' (in Frege's terms) in that it evokes something of the evaluative attitude one takes toward it as a particular organism. Human languages are replete with such resonances.

Marsden comments on the philosophy of discourse as it applies to Māori people.

> The route to Maoritanga through abstract interpretation as a dead end. The way can only lie through a passionate, subjective approach. That is more likely to lead to a goal. As a person brought up within the culture, who has absorbed the values and attitudes of the Māori, my approach to Māori things is largely subjective. The charge of lacking objectivity does not concern me; the so-called objectivity some insist upon is simply a form of arid abstract action, a model or a map. It is not the same thing as a taste of reality. (2003, p. 2)

He concludes his thoughts on being Māori with the thought:

> The integration of an individual into full membership of society take place over time, not in formal school, but in his living situation. The process of learning by which the raw material of the young is transformed into full

> citizenship is inherent in the workings of each institution so that the install-
> ing of values, existence is in the cultural menu. (2003, p. 23)

The idea of apprenticeship is allied to seeing language as a complex of techniques. This fits with the norms of everyday life. Thus, an individual is not so much explicitly instructed but inducted into ways of going on. The learning involved encompasses the possibility of both social morality and individual ideal, to use words borrowed from the philosophical reflections of Peter Strawson (1961). He makes the point that human society in general creates in recognised ways (e.g., artists and poets) according to norms. These norms also must be woven into the joint life of members in community through close relationships such as family or tribe. They may include even wider relationships such as those adopted when a group recognises norms to be so important that they enact laws that bind them all. These are, to some extent, loose and open to variable interpretation, as the recent conflict between Russia and the Ukraine has shown. However, even with these socio-political norms, we might strive as human beings to uphold and defend some universal goods.

Among the loose norms which allow human life to flourish, there are a set of 'reactive attitudes' (Strawson, 1974) or ways of regarding each other which take note of our human vulnerabilities and needs and show sensibility to the human condition. These can reflect adversely on seemingly privileged actions, such as those of a government, to the point where we recognise international tribunals in an attempt to curb politically motivated and endorsed violence as a form of crime—war crime. These disputes are examined and resolved between nations within broadly agreed ways of going on. Certain crimes in most societies are recognised as requiring correction so that norms are upheld such as the non-exploitation of children, the prevention of harm to the innocent, consideration for the infirm or suffering among us, and so on. In war, the widespread slaughter of civilian non-combatants is also generally regarded as a crime against humanity.

Natural Language

Julia Kristeva, remarking on the idea of natural language and speech as involving norms which apply to us all, notes that it is organized like an immense combinatory, like a universal calculus laden with mythological, moral and social values; speech, however, does not isolate the act of signify in the act off signifying—its verb—in a mental elsewhere. The participation of language in the world, in nature, in the body, in society ... constitutes perhaps the fundamental trait of the conception of language in so-called 'primitive' societies'. (Kristeva, 1989, p. 4)

This is a clear statement of the naturalism common to post-structural thinkers. It places them in broad continuity with the later, more sophisticated views of Wittgenstein on thought, language, and meaning. He radically not only reformulates but, in many ways, rejects any *Tractatus*-like view that "the facts in logical space are the world" (1.13).

According to his more anthropological and nuanced view, we have discursive ways of interacting and diverse aspects to our many ways of going on together. In this way, our adaptation to 'the slings and arrows of outrageous fortune' can be cooperative and supportive, giving us tremendous advantages over creatures whose lives are 'nasty, brutish and short', even if variably communal.

These can also be seen in sport where discourse and playing the game are woven together so that one cannot coach, or even intelligently enjoy, a sport without using discourse especially developed for that sport. The incorporation of discursive elements into the shaping of a sporting performance marks the difference between a novice and an elite player. In cricket, it is when one learns not just to bowl the ball but to be able to bowl a delivery on the 'leg stump' (or just outside) which will spin back and hit the wickets that one becomes a 'leg-spinner' worthy of the name. Any game is a human performance with many aspects, some of which are subjective and incline us to act in a certain way, such as the batsmen lunging toward the side of the pitch opposite to that on which he is standing relative to the wickets. That very subjective if understandable tendency is made to look inept when the ball grips on the surface of the wicket (or pitch or playing surface) as it bounces and turns past the bat, so that the batsman sees the wickets hit behind him. He is then 'out'. His turn at bat or 'innings' is over.

Here we return again to the subject of linguistic meaning and notice that there is a heavily cultural aspect to meaning. This aspect imbues even a broadly Fregean interpretation of 'reference', 'denotation', or 'sense'. Reference is straightforward and may be functionally specified, though certain terms create difficulties. 'Blue', for instance in a similar way to 'pink', can denote a colour, a mood, a style of music, a political persuasion, or a type of literature, to select just some of the areas of discourse that might be involved. Even a denoting term such as a proper name can carry a range of senses, some of which merge into emotions and subjective feelings or clouds of related responses in diverse human beings or situations (Gillett, 1992, p. 140, 151ff.). For example, one expects a juicy-looking green apple to be crunchy by its visual presentation. Yet it may in fact be rotten at the core. That misapprehension is entirely understandable given the diffuse and holistic nature of the associative connections in the human neural network and normal ecological circumstances.

As noted, the neural system is self-modifying according to the many activities and discursive encounters of a human individual with multiple others in a context. Those others have their own discursive networks which influence their perceptions, cognition, emotions, and interactions with the world and with others. Thus, each person's and our shared behaviour and orientation to events is constantly being updated and triangulated. This takes place against the view of a notional 'third person'. It makes any neurophilosophy framed in terms of the inputs, intermediate states and events, and outputs of a single located and articulated or surveyable system a gross oversimplification of the way discourse changes us individually and collectively as we participate in different social contexts.

In a telling article mentioned earlier in this chapter, Peter Strawson considers the many allegiances and associations that human beings have in the service of both their interests and any common goals or enterprises they might pursue (Strawson, 1961). He notices these diverse appeals and demands on us, some formal such as the law, others informal such as a neighbourhood group in which folk normally help one another. These may mean that they share fruit from their properties, given that different plots of land may favour growing different types of plant and between them they may, for a limited season, greatly enhance each other's lives in a diversity of ways. Such arrangements, ranging in varying degree of complexity and formality, include both the informal and spontaneous events people share in neighbourhoods and the constitutions of states. The formality of the laws of a nation by which people live and organise their lives together in a society build on and systematise these human relationships and agreements. They pervade the human world in myriad ways and are facilitated by human discourse. The resulting phenomena draw on both reactive attitudes to different behaviours (Strawson, 1974) and systematic (or quasi-systematic) institutions like natural language, at the one end, and

explicit constitutions and treaties, at the other. Thus, we live with normative constraints and opportunities which go far beyond those offered by
our highly evolved biological nature.

This is the world of *l'etre et le neant* explored by Jean Paul Sartre and
his many social and academic interlocuters. That human world is 'being'
in its full ethological complexity and also draws on the 'nothingness' of
human imagination and the possibility of human relationships through
history, current society, and the future which we approach together.

Kristeva ties together language subjectivity and 'persons' and interro-
gates 'tense' as a discursive device related to time and grammar (1989,
p. 32). She also notes the multiple variations that words can exhibit in
relation to what they are about and therefore in their 'meanings' as they
pervade 'the life of the mind' as we participate in it together (p. 38). In
discussing the human psyche, she considers language as a key to conscious
and unconscious aspects of human life, seeing 'speech and reply' as basic
to all human life (p. 266). She writes of enabling 'science to grasp the
multiplicity of signifying systems elaborated in and from *la langue*'
(p. 277). This perspective takes us a long way from quasi-mathematical
approaches to language, and we again see the complexities that lay beyond
and failed to be recognized in the *Tractatus*.

A discourse that seems as if it should particularly support the logico-
mathematical turn taken by Frege is mathematical language. Mathematics
has for many seemed to hold the promise of rendering philosophy in a
systematic form apt for articulation in just the way Frege and the neo-
Fregeans envisaged. But this is not how our neurocognitive system processes language and constructs meaning from it. Wittgenstein famously
undercuts the logico-mathematical position in his own *Remarks on the
Foundations of Mathematics* (1983). Although he did not discuss our neurocognitive system as we have, his comments support our rejection of the
view that complex equations can explain that this system is mathematically
determined by appropriate stimuli to operate in certain ways.

The steps are determined by the formula... (#1)

But then what does the peculiar inexorability of mathematics consist in?—
Would not the inexorability with which two follows one and three two be a
good example? But presumably this means follows in the order of the cardi-
nal numbers; for in a different series something different follows. And isn't
this series defined by this sequence? (#4)

> But if you were right, how does it come about that all men (or at least all normal men) except these patterns as proofs of these propositions?—It is true, there is a great—and interesting—agreement here. (35)

Wittgenstein goes on to thoroughly embed mathematics in our practices. This involves not only dealing with everyday tasks, such as counting apples, but also instructing our young to do things our way and so function smoothly as part of our community who 'do things with words' (like the builder and his labourer of the *Investigations'* #2). We are creatures who learn to work together. Wittgenstein systematically considers rule-following, calculating, the role of experience, propositions, irrational numbers, and mathematical proof as a game or a set of techniques. These techniques are related in the same way to the world as any other technique such as art or, even more closely, toolmaking. This technique in particular can be stylised and turned into an art form as, for instance, the aesthetics of prehistoric stone blades (Currie & Zhu, 2021). We can link aesthetic creations to everyday skills used to deal with the environment and honed to a degree of attainment that has lasting effects on the skill-using group to their ultimate ethological advantage.

Thus, mathematics, that most articulated and seemingly formal or determinate variety of cognition, takes its place as a language game among others with rules of procedure which guide one through an area of shared life in the world. The artwork of prehistoric stone blades, on reflection, seems to celebrate the human condition and human ability to extend beyond the functional into the aesthetic realm. These and other productions have a supplementary but nonetheless important role in practices of value and exchange. They subserve both connectional and conventional roles in human society.

From Abstraction to Practice

Our children are apprentices in this activity. As one attains maturity, a further development of this very human orientation to life occurs. We become fully participating members in socio-political organisations governed by ethos and, more formally, the codified institution that is the law. The law codifies principles or higher-level abstractions from the intuitive relationships that allow us to function as a successful collective. Like all abstractions, including logic, engagement with the real world of human life must be retained and constantly assessed. That engagement can be made and improved against an abstract ideal. What results is a complex system replete with jurisprudence. That aspect of law, in many cultural contexts, involves the presentation of one's reasons and motivations for one's actions before a group of peers who bring with them the ordinary intuitions and attitudes of the community and not the abstract knowledge of learned professionals. They add what Frege would have called 'a pinch of salt' to the cognitive recipe of inference and evaluation and so reinstate the intrinsic and inclusive intelligence of ordinary human beings into the judgement. (Some of us are more aware than others of the silliness into which an exaggerated tendency to abstraction can lead us.)

Thus, we have come full circle. We have moved from all-in consideration of flexible being as an aspect of human ethological adaptation, to a set of philosophical abstractions, and then to a brief conclusion about jurisprudence and the human face of the law as an adaptive tool given conflicts and uncertainties within a human group.

The law is a reflective and normative abstraction which serves as a useful tool to regulate our complex human socio-political life. As such, it is part of the constantly updating and self-reflective work of human cognition. It is normative and, as human norms of various kinds and degrees of an abstraction and formality criss-cross our lives as neurocognitive beings, we build on its demands and complexities in many ways. Appreciating these demands and complexities as processes that go beyond abstract formalities is a cognitive task with many ramifications in our shared lives. That is evident in the many branches of law: property law, criminal law, constitutional law, traffic law, international law, international criminal law, the laws of war, and so on. Each of those is complex in its own right, intersecting in various ways and entangled with our ordinarily ways of going on as human beings and human societies in the world.

In fact, when we examine the law from a naturalistic philosophical view, the law itself can be seen to embody correctives to over-formalisation and the risk of losing touch with human life as it is lived. There is, for instance, the widespread practice of trial by jury whereby one's case is presented before a jury of one's peers in order to have the dispute seen from the point of view of the everyday 'man on the Clapham omnibus' (or 'woman on a flower-stall'). Not only does this face of the law attempt to represent the realities of a case from a point of view that an ordinary person could understand. It also tries to ensure that the judgement involved has that same worldly understanding incorporating *Menschenkenntnis*. Jurisprudence is an aspect of the law that interprets statutes and what they should safeguard and represent in realistic and everyday terms. It interprets them as they would be understood and attempted to be followed by an ordinarily informed member of society imbed with its moral norms and considerations.

Taken together, the many shared practices and contexts in which we act tend to mean that the actions of a human individual and member of a community are not framed and judged by a privileged elite or academic in possession of uncommon knowledge (and perhaps an autistic frame of mind). Rather, they are framed and judged by peers or social equals living with the same constraints and moral expectations as anyone accused of a crime or subject to public scrutiny, though that can be manipulated discursively.

In all societies, including indigenous societies, some collective such as a tribal council or other body of fellow human beings are charged with dispensing justice in disputes. Those who have a special knowledge of tribal or community traditions and spirituality may be involved in such a reflective activity. The latter ensures that even those entrusted and privileged with special roles in this process hold themselves accountable to a higher authority who might be considered all-knowing or at least cognisant of more eternal and lasting values. Of course, because, in reality, these are human beings subject to all the distortions of cognition to which 'flesh' is prone, such authority usually carries considerable responsibility and answerability to the community.

The traditional Platonic doctrine of the immortality of the soul is pertinent here as we consider theology and the law in post-enlightenment societies (Broadie, 2001; Karasmanis, 2006). Whereas Plato was open to religious and traditional beliefs, Aristotle sought the answers to questions about human nature on the basis of naturalistic philosophy (Charles, 2021). This was a difficult task in the day, as there were no developed sciences of neurocognition. We are now armed with dynamic neurophilosophy as a recent iteration of such scientific knowledge, and it attempts to shed light on the way in which the brain instantiates in its neuronal circuitry the rhythms of life.

That development is a game-changer in the post-enlightenment world as it simultaneously allows us to adopt a naturalistic stance while not being reductively functionalist in neurophilosophy. We can replace mysticism or agnosticism about the adequacy of science to understand human neurocognition with a more thoroughgoing naturalism which does not attempt to force nature into the Procrustean bed offered to it by Victorian

attitudes to natural abilities. Dynamic neuroscience transcends program-
matic metaphors based on human invention and allows for the fluidity of
genuine biological embodiment. Following continental philosophers, we
have called this 'being-in-the-world'. We can call it as such without retreat-
ing into a precious mode of anti-scientific philosophising.

If that orientation can illuminate certain aspects of our spirituality, so
much the better. That is indeed a realistic hope with the idea that the communicating minds and shared lives of human beings confer on any human
being a kind of profile beyond his/her biological life. Our names are per-
petuated by our familial and cultural naming practices. Along with our
names go various celebrations and commemorations of our lives which
extend our identities beyond our lifetimes. Our identities are not just biological and psychological but biographical as well. These, in turn, can be
viewed *sub speciae aeternitatis*, and so against a more inclusive framework
than that of our own socio-political contexts. Given that such a shared and
growing presence in the minds of other human beings is potentially unlim-
ited, a kind of immortality and universality attaches to any of us, whether
that is a pleasant or unpleasant thought, the inner life of self can be either
'heavenly' or 'hellish'.

On a more down-to-earth note, we can consider the place of disabled
human beings amidst a society of 'normally-abled' peers. A short-sighted
tendency is to see the disabled as a burden. But here we can learn from
indigenous societies and their regard for the elderly. The elderly are not
necessarily disabled or unfit. They may be less fit to meet current environmental challenges than the young, especially when we consider changes in
socio-political institutions and technology. But in place of physical ability
and adaptability they are, jointly and severally, a repository of group memory going beyond individual lifespans. This is a resource that we in the
post-industrial West tend to call 'institutional memory'. It is otherwise
denoted by the term 'tribal memory' or the more contemporary scholarly
term 'cultural memory'.

Even in the developed West, this type of memory has a special place.
Who does not remember a scattering of Greek myths, Norse sagas, or
Anglo-Saxon stories the origins of which are clouded by the mists of time,
for example, in Beowulf? More modern historical fictions draw richly on
this treasure trove of shared memory. We need think only of 'Hamlet' or
'MacBeth' to name two famous examples whose origins are obscure and
whose provenance is through the art and literature of British culture.

Philosophers of neurocognition tend to refer to counterfactual thinking and wonder about its provenance and role in human discourse. Shared cultural memory does include counterfactual thinking but also includes art and the fantasies we enjoy from childhood through youth and through into adulthood where the genres of historical and fantasy fiction are well represented in booksellers' lists. We all, to varying degrees, enjoy a good story, some inclining to the more historical or realistic and some to the more fanciful and imaginative, but nearly all requiring a certain semblance of believable human reality.

It is surely part of our epigenetic store of techniques that fiction of various varieties, commends itself to various of us just as disabilities of various kinds carry an advantage to the group who makes them part of its flexible adaptation to the world. We thrive, as a species, on flexibility and innovation. So, we should make them an enjoyable part of our rhythms of life, nurtured in childhood and then enjoyed for its own sake as adults who find different aspects of childhood and being part of the human ontogenetic development involved. We find this rewarding and engaging to various degrees. Such variations often complement one another so that one might find a married couple in which a visionary is allied to a capable and socially adept citizen of the world. We even make jokes about such matters:

'I leave the big decisions to my husband', remarks a woman, horrifying her feminist friends.

'How can you?' they retort, chiding her for her sexist and reactionary stance.

'Oh easily' is her response, "He decides whether there is a God, did the universe have a beginning, is it possible to think beyond infinity, and when will the world end. I just take care of everyday details such as 'where we are going to live,' 'what job he will do', 'how we spend our money', and 'who our friends will be'; it works very well with me in that role".

Despite being mildly amusing, the joke makes the useful point that our thinking is of various kinds and that counterfactual, imaginary, and in other ways non-applied content, is part of a very human characteristic (perhaps with gendered variations).

This inclusive and varied construal of thinking and its role in our being does not apply in the same way to animals. We say this despite the fact that we may remember our pets in ways replete with human sentiments attached to those memories and relate to them in ways influenced by many anthropomorphisms. We may even remember cars or houses with certain feeling so that there is a difference between the cognitive profile of 'a house' versus 'a home', and our children often 'sense' or even 'feel' those differences deeply and in an inescapable way. Being human, we might say, with all its resonances, is always with us and part of our being in the world.

CONCLUSION

Our journey from nature (unadorned and thought of as involving mainly 'tooth and claw'), has led so far to discourse and all its complexities and philosophical abstractions in which language has figured large. We shall have occasion to return to language as we consider discourse more widely and locate it in its proper ethological context in the remaining two chapters.

At this point the rhythms of recovery, as in the case of Nick Chisholm, become important. In recovery one seeks to recover function, which is better thought of as the ways of going on or holistic neurocognitive rhythms that make one 'the person that one is'. Some will have been rendered unattainable because of one's disabilities or morbidity but, given imagination and empathy, paths to recovery may be found. This discovery is contingent upon the person concerned coming into view within the therapeutic gaze as is well appreciated in occupational therapy. A recent analysis of dominant features of practice identified contextual factors, personal challenges, and socio-cultural challenges in the context of a conducive relationship allowing the client/patient to develop and realise goals that they found meaningful within an inclusive understanding of their life as the person concerned (Murray et al., 2015).

An earlier study of indigenous clients in Australia had identified similar factors and specifically focussed on the need to be client-centred and for therapists to have cultural awareness and socioeconomic sensibility in order to appreciate the life challenges and goals of clients who are also 'patients' in that they suffer though not straightforwardly from recognised diseases and pathologies (Stedman & Thomas, 2011). The authors note an ongoing need for self-reflection by the therapist, which adds an important inter-cultural aspect to the therapeutic relationship particularly in post-colonial settings and highlights an oft-neglected aspect of the bio-psycho-social model of health and disease (Bolton & Gillett, 2019).

Although the connection is not of any explanatory or interventional use, a recent survey of the interwoven and complex biological changes of aging is salutary in reminding us of the difficulties and limited prospects for any intervention that might significantly affect the human life span (Milholland & Vijg, 2022). One is thrust back upon complexity and chaos theory and the need to understand our human and mortal limitations as neurocognitive organisms in this world. This poses many challenges and salient questions to us which, in many guises, cause us to reflect upon ourselves and the many dimensions of our being. Each dimension structures properties which are, in nature as in human life, deeply entangled and spur different reflective paths through the maze of human experience and learning and cannot be 'straightjacketed' to fit a Procrustean bed of academic merit (apologies for the mixed metaphor). We must therefore learn to deal with complex reflection of vastly different provenance from biomedicine to history and culture in order to sketch the knowledge deeply embedded in any discussion of the human soul. In order to address key issues in that reflection, we shall have to re-immerse ourselves in the debates around consciousness, discourse, and intention.

References

Bolton, D., & Gillett, G. (2019). *The biopsychosocial model of health and disease: New philosophical and scientific developments.* Palgrave Macmillan.

Broadie, S. (2001). Soul and body in Plato and Descartes. *Proceedings of the Aristotelian Society, 101,* 295–308.

Charles, D. (2021). *The undivided self: Aristotle on the mind-body problem.* Oxford University Press.

Chisholm, N., & Gillett, G. (2005). The patient's journey: Living with locked-in syndrome. *BMJ, 331,* 94. https://doi.org/10.1136/bmj331.7508.94

Currie, G., & Zhu, X. (2021). Aesthetic sense and social cognition: A story from the early stone age. *Synthese, 198,* 6553–6572.

Eccles, J. (1970). The brain and the unity of conscious experience. In *Facing reality: Philosophical adventures of a brain scientist* (pp. 63–84). Springer.

Fogal, D., Harris, D., & Moss, M. (Eds.). (2018). *New work on speech acts.* Oxford University Press.

Gillett, G. (1986). Brain bisection and personal identity. *Mind, 95,* 224–229.

Gillett, G. (1992). *Representation, meaning and thought.* Clarendon Press.

Halan, T., Ortiz, J., Reddy, D., et al. (2021). Locked-in syndrome: A systematic revie of long-term management and progress. *Cureus, 13,* e16727. https://doi.org/10.7759/cureus.16787

Karasmanis, V. (2006). Soul and body in Plato. *International Congress Series, 1286,* 1–6.

Kristeva, J. (1989). *Language: The unknown: An initiation into linguistics.* Columbia University Press.

Luria, A. R. (1987). *The man with a shattered world: The history of a brain wound* (L. Solotaroff, Trans.). Harvard University Press.

Lycan, W. (2019). *Philosophy of language: A contemporary introduction* (3rd ed.). Routledge.

MacIntyre, A. (1989). *Against the self-images of the age: Essays on ideology and philosophy.* Notre Dame University Press.

Marsden, M. (2003). *The woven universe: Selected writings of Rev. Māori Marsden* (T. A. C. Royal, Ed.). Estate of Māori Marsden.

Milholland, B., & Vijg, J. (2022). Why Gilgamesh failed: The mechanistic basis of the limits to human lifespan. *Nature Aging, 2,* 878–884.

Murray, C., Turpin, M., Edwards, I., & Jones, M. (2015). A qualitative meta-synthesis about challenges experienced in occupational therapy practice. *British Journal of Occupational Therapy, 78,* 534–546.

Nagel, T. (1971). Brain bisection and the unity of consciousness. *Synthese, 22,* 396–413.

Parfit, D. (1984). *Reasons and persons.* Clarendon Press.

Schwartz, J., Stapp, H., & Beauregard, M. (2005). Quantum physics in neuroscience and psychology: A neurophysical model of the mind-brain relation. *Philosophical Transactions of the Royal Society of London, B: Biological Sciences, 360,* 1309–1327.

Searle, J. (1969). *Speech acts: An essay in the philosophy of language.* Cambridge University Press.

Shaffer, J. (1977). Personal identity: The implications of brain bisection and brain transplants. *Journal of Medicine and Philosophy, 2,* 147–161.

Southwick, S., Charney, D., & DePierro, J. (2023). *Resilience: The science of mastering life's greatest challenges,* third edition. Cambridge University Press.

Stalmaszcyzyk, P. (Ed.). (2021). *The Cambridge handbook of the philosophy of language.* Cambridge University Press.

Stedman, A., & Thomas, Y. (2011). Reflecting on our effectiveness: Occupational therapy interventions with indigenous clients. *Australian Occupational Therapy Journal, 58,* 43–49.

Strawson, P. F. (1961). Social morality and individual ideal. *Philosophy, 36,* 1–17.

Strawson, P. F. (1974). *Freedom and resentment and other essays.* Methuen.

Tarlaci, S., & Pregnolato, M. (2016). Quantum neurophysics: From non-living matter to quantum neurobiology and psychopathology. *International Journal of Psychophysiology, 103,* 161–173.

Wilkes, K. (1993). *Real people: Personal identity without thought experiments.* Oxford University Press.

CHAPTER 6

Consciousness, Discourse, and Intention

The entrée into consciousness and intention is through discourse. Yet the idea of qualia, or 'raw feels', seems to direct us away from that route. The descriptor 'raw' sends thought in a more grounded direction and reconnects with the prehistory that encompasses neurocognition as an anthropological moment or transition from one stage of thought to another. We are recently reminded of this aspect of human evolution in life. It is clearly epigenetic in the wide sense we have outlined. The *Uhuru statement from the heart,* produced by indigenous folk in Australia, clearly expresses postcolonial grievances (indeed atrocities including genocide), and it is affirmed a healing recognition and reconciliation in response to this and other forms of social harm. This holistic and inclusive discourse transcends such human evils towards healing. Discourse is thus one way to overcome our emotional deficiencies. We can use it to graciously forgive, while being honest about and confronting those who have wronged us, thereby urging reflection in a spirit of repentance so as to allow us to move on in a constructive way.

With its complexity and fluid subtlety, discourse therefore evades quasi-mathematical or formal straightjacketing, much to the chagrin of a certain scientific way of thinking. When we realise our limitations in this regard, we are prepared for what Wittgenstein thought of as 'the hard work' of philosophy. Here we might follow certain leads suggested elsewhere and explore the relations between language, structure, and the unconscious.

G. Gillett, W. Glannon, *The Neurodynamic Soul,* New Directions in Philosophy and Cognitive Science, https://doi.org/10.1007/978-3-031-44951-2_6

The semantically rich term 'signification' hints at the role of language in linking many descriptive and communicative purposes. These can be loosely gathered under Wittgenstein's term 'grammar' so that an overview of their role in that ramifying structure could be examined to disclose the psychic complexity of language. As we noted in Chap. 5, Kristeva is also interested in the structure of language from a dual entrée via both linguistics through the 'network of signifiers' (1989, p. 20) and 'psychic structure' (p. 213). It is tempting to accept appealing syntheses in this complex area. An assimilation between signification and the structure of reference and sense springs all too easily to mind, but on the basis of Frege's hints and the limitations of early Wittgenstein (particularly *Tractatus*, 6.52ff), we 'eat the air'. We can thereby seem to neglect our complex entanglement through psychologically rich forms of life with a real world and other people to whom we are related both affectively and actually. In this way, our explicit thoughts abstract necessarily and in real life multiply from those engagements (Gillett, 2001). The relevant resonances are very rich indeed, in ways that the later Wittgenstein came to see (Read, 2001) and, in so doing accepted the anthropological and affective or relational complexities to which we have referred.

Read wishes to return language to its primary role in the real world. That is fraught with the philosophical hazard posed by a "causal naturalism undermining human 'autopoiesis'" (McConnell & Gillett, 2005, p. 63). This brings us to Freud. He was a Darwinian causal determinist and carried that metaphysics into his writing on culture and human relations along with his admixture of actual psychology and an ever-present concern with sexuality and incest taboos.

A return to situated and ethological *praxis* and culture avoids any latent relativism or idealism and forces us to 'get real'. That is nested in our embodied and embedded existence in the real world. But it defies persuasive scientific abstractions based on interpretations fashionable in some context or among some complex of self-images with its own history and cultural setting.

Wittgenstein was both fascinated by and critical of Freud and psychoanalysis (Wittgenstein, 1967). He noted Freud's emphasis on 'causality' despite the lack of experiments and "causal laws'" (p. 42). He also reflects upon Freud's writing on dreams, relating them to symbols, art, and dream messages (p. 45). He later invokes 'games and children's play', thereby loosening the mind's activity from rules so that theories of play invoking causes generally go wrong (p. 49). He advises a critical stance towards theories in psychology especially when they rely on creating a causal account out of the thought "Yes, of course, it must be like that"(p. 52).

NEURODYNAMICS AND THE UNCONSCIOUS

Neurodynamics allows us to think again about Freud and the unconscious mind rather than creating "A powerful mythology"(Wittgenstein, 1967, p. 52). The mammalian brain develops holistic rhythms weaving together sensorimotor and cognitive activity to enable adaptation via the development of practiced response patterns in an ecological context. These patterns are developed through a combination of inheritance, developmental learning, fitting in to group patterns of activity, cultural assimilation in the human case, and discursive 'shaping'. The result has a certain style of being which involves emotional resonance, arts and crafts, patterns of family, and community life. These alter our neurodynamic resonances and inform a broad range of behaviours.

When we abandon or loosen and modify an overly formal and therefore misleading view of the structure of language and an allied 'picture theory', we can rediscover the idea of life as it is lived and language as a sociocultural tool that is constantly changing. In that *Übersicht*, our lives resemble 'braided rivers' coursing through a partly human and partly prehistoric landscape, as they do in The South Island of New Zealand (Macfarlane et al., 2015; Macfarlane & Macfarlane, 2019). Human life and natural language is partly the product, but also forms the currents and the riverbed and thus the flow and sounds of the phenomenon in which we are immersed and which we are considering. It has depths barely discernible from the surface but known to those who study and reflect on and fish in it. It has currents we feel and understand as we enter into it and swim in it, particularly if we then reflect on that experience. Reflection on the forces acting on us during our immersion allows us to become conscious of, and cognitively engaged with, aspects of the forces acting on and within us. This reflection can be pleasant, painful, provoking, or challenging but seldom leaves us unaffected. At best we learn, about ourselves, our relationships, human life, and its complexities, indeed sometimes about many things that are entangled or intersecting.

Taken as a complex confrontation and sometimes an insight into ourselves, we are changed from the reflective process. The change is seldom instant or dramatic. Yet all traditions recognise such a possibility, and some of us, like the character in the story, can say 'Do I believe in it; hell,

I seen it!' (where 'it' refers to marriage.). Moreover, pun, as a psycho-linguistic phenomenon, depends on knowledge which lies unattended to until 'shaken or stirred' (with apologies to 'James Bond'). Freud's writing resonates with many such confluences in the human neural nets of broad cultural groups under the binding and encompassing *aegis* of Darwinian theory. We the authors aim to be broadly of that tribe if somewhat errant (as in knight errant) or perhaps enterprising, or even wayward, members.

THE ROOTS OF OUR PSYCHOLOGY

In common with Sigmund Freud and his approach to psychology (or more accurately Hughlings-Jackson), we would trace many psychological phenomena to their developmental roots, acknowledging Vygotsky's important work on language and culture. This work immerses us in the psychology of the child with its affective attachments and dependencies and its approximations and imitations of adult conduct and speech. Regional accents make the point more vividly than abstract scholarship. To successfully imitate an accent is, in part, to get into a mind set in which the world itself seems subtly different. As adults, we leave behind many of these childhood habits and obsessions and moderate others. There are, after all, distinguished academics among the many adults who enjoy 'war-games' in one form or another and even spend hours obsessively painting model soldiers. Childish things can fascinate and absorb us, and sport or other forms of play remain a prominent human activity in all cultures. In fact, dreams and desires, some all absorbing, pervade sport at every level and sometimes redeem the children of the poor whose sporting prowess can bring them lives, societal recognition, wealth, and social positions of which they could otherwise only dream.

The realities of our unspoken minds are a fertile ground for creative thinking in art, psychology, and religion. Wittgenstein's reflections in these related areas of the mind are perceptively gathered together in one volume and reveal the confluences between them (Wittgenstein, 1967). Wittgenstein remarks, "The subject [aesthetics] is very big and entirely misunderstood as far as I can see"(p, 1). This remark could equally apply to certain forms of psychology (at least that psychology—deeply influenced by Freud—on which he focused) and religion. All three, we suggest, are revealing manifestations of the human soul. In aesthetics, Wittgenstein contends that we project our own reactions, both favourable and unfavourable, on artistic objects and initiatives and that our aesthetic words and judgements are not merely adjectival but embedded in every-day experience and reactions (p. 3). In some ways, they are of a piece with our emotive responses as a dog might wag its tail (p. 6); but they span a broad relation to their topic discourse with its own complexities. Aesthetic reactions and thoughts have a complex place in our neurocognitive lives related to memory, associative discourse, (Wittgenstein, 1967, p. 19) and so many other facets of experience that the psychological phenomenon of '*freier Einfall*' becomes relevant (p. 25). Aesthetic puzzles (p. 28) weave fascinating, occasionally musing, and serendipitous paths through the human neural network catching our minds in its own 'wonderland'.

Into this evocative milieu step Jacques Lacan (1977) and Julia Kristeva (1989). They enter in the spirit of Freud and his powers of weaving together aspects of the psyche that fascinate and enthral us. Out of the triad of culture, truth, and science, such thinkers brew a heady mix in which the human psyche contrives its own seduction (Gillett, 2015). Our human embeddedness in *praxes* and ways of knowing spawn a cultural construct, part of which are our related conceptions of 'truth' and 'real-ity'. Although our interwoven lives draw on multiple subjectivities, they all take place and arise within an entanglement with the world in which con-tingencies shape our individual and collective souls so that each soul or psyche shares a mix of commonalities and individual characteristics indi-viduated by a name (which, as we have noted, itself 'speaks' of family and culture). Thus, we are made and not just born in ways that Freud and Kristeva both invoke in their respective syntheses.

Lacan, with his distinctive post-structuralist synthesis, contributes to this heady work and can net a rich haul of academic attention by drawing on the ever-intriguing Freud. This focus includes an enduring fascination with biblical illusion, structuralism, and the hotbed of ideas that is Parisian academia. Our voice in this is but a modest, and hopefully corrective cri-tique in the light of contemporary neurodynamics. The Wittgensteinian spirit should be evident but is, of course, merely our take on the work of Freud and Lacan.

When one allows a sense of critique that is neither wholly critical nor dismissive, we come close in spirit to Wittgenstein's view of religion.

> If someone said: "Wittgenstein, do you believe in this?" I'd say: "No." 'Do you contradict the man?' I'd say "No." (1967, p. 53)

He goes on to dissociate the question from 'belief' in the ordinary epistemic and propositional sense and eschews the idea of reasons, evidence, and rationality as we understand them in philosophy (pp. 56ff). In reflecting on 'God', he muses on our learning the word, our reports of mystical phenomena, miracles, and folkways in evaluating sacred words, and he adverts to the picture theory in appreciating 'higher things' noting great human achievements such as Michelangelo's art (p. 63). This is not a dismissive attitude but closer to what we now call 'deconstructive', with a generous infusion of respect and appreciation for the origin and discussion of the term. He moves to a conclusion, such as there is, with the remark "The whole *weight* may be in the picture"(p. 72).

Human beings have developed evaluative discourses to assess our reactions, techniques, and languages of interaction with the world and each other. These discourses eschew mere elaboration and descriptive predication. The use of 'evaluative' here pushed us more towards 'affective' than 'normative'. The relevant techniques are not able to be prised apart from our ways of going on. Instead, they allow us to reflect upon our lives, using the full resources of our holistic neurocognitive system. That system therefore ranges over detailed and nuanced techniques of perception, action, and cognition (Gillett, 1992), replete with cultural resonances and artefacts, some of them discursive. Wittgenstein withheld committed belief to any religion but did not allow that to cause him to disrespect them (Malcolm, 1965). It was a wise and sensitive stance on his part. In allowing us to explore our language and its intricacies, what Wittgenstein calls 'grammar', becomes part of understanding human adaptation and cognition. An important aspect of our cognition is human imagination and thoughts about our place in the world and what we call 'goodness' or thriving individually and collectively. That involves both practical and moral reasoning that evaluatively influences our ways of going on and the rhythms of life in which we are engaged.

The authors have previously deconstructed the question of free will by offering a neurodynamic account of autopoiesis. The frontal lobes of the brain engage holistic resources—both inside and outside the brain—to deliver a variously individual response to different situations in which a human being is involved (Gillett & Glannon, 2020). This is one aspect of how our neurocognition extends into the general context of action.

CONSCIOUS AND UNCONSCIOUS PROCESSES

The frontal lobes, as we have noted, communicate with wide areas of the cortex and have top-down influences on many subcortical areas of the brain, giving and receiving extensive connections to and from them. Thus, the pervasive influence of frontal lobe activity has evolved to incorporate more and more advanced effects in the increasingly complex array of neurocognitive adaptations we have developed.

Wittgenstein's attempt to ground the main questions of philosophy in natural language as it is ordinarily used could be construed as reviving the idea that 'metaphysics' is that which comes 'after physics' as the original scholars giving us the Aristotelian corpus rendered it. That would make his contributions particularly apt in exploring the primacy of neuro-philosophy as a key to philosophical reflection. We can now elaborate how conscious and unconscious processes influence different aspects of human thought and behaviour.

Inhibition and Control

Top-down inhibition related to conditions in which an action has adverse consequences comes into action in diverse situations. These range from regressive tendencies associated with pain and the presence of nociceptive stimuli to an animal's negative response to withdrawal from conditions associated with positive stimuli or a frequent species-typical pattern of responding. Animals have fine and malleable discriminative abilities. These include signalling the whiff of a truffle in a suitably trained animal or tracking an engulfed human being as in a snow rescue dog. We exploit these tendencies in various animals and become attuned to very complex sets of conditions in our own skilled activity, such as hunting or tracking. We develop complex procedures of sensorimotor response and social coordination to use collective capabilities in such conditions. The relevant capacities are embedded in discourse and reinforced both positively and negatively by discursive interactions. Our children enter into this complex *milieu* and emerge from infancy with varying degrees of mastery of the techniques associated with life among others in the world.

Natural Language

Natural language is a means of a communicating cultural group conveying thoughts to each other replete with the group patterns of behaviour and sensibility that better equip them in an ethological context. Its discernible structure can create the impression of a quasi-mathematical essence to this human phenomenon. Such abstractions were critiqued, as we have noted, by Wittgenstein in his move from the *Tractatus* to his later views. Real language, like any other ethological feature of human life, grows organically and over time and approximates then realises its extant dynamic form in a given society at a given point of history. Philosophers and linguists then 'hyper-reify' its quasi-mathematical features into the structure of language as Frege did, followed by the young mathematically inspired Wittgenstein with his 'philosophy in its final and perfect form', as he claimed in the *Tractatus*.

Social Reward and Human Infancy

Learning not to be distracted and yet to remain open to significant events is a skill which a child gradually masters. We all, as children, progressively learn the open-ended and creative interweaving of life and language. We have noted that neuroscience has been through a number of fashions in its thinking: from dualism, to mechanism, to quantum concepts, and now AI and our biological dynamism. No doubt there will be more, and the best we can do is offer a contemporary synthesis respectful of various traditions and openly aware like 'Dirty Harry' (suitably gender-adjusted); "A person's gotta know their limitations". Language, through its interpersonal emotive resonance is a powerful shaping influence on thinking and thereby thought in the holistic, commonly conceived, and non-Frege-Russell logico-mathematical sense.

As people mature, they learn to move beyond their childish and adolescent preoccupations. These are useful in training us to develop skills which will complement others in our group—extending existing skills in new ways and opening up new areas with new possibilities. We do this with the intensity of youth, and thereby the group has members who are innovating in ways that will benefit all the group as a micro-culture. It also has members that are more experienced or measured, so less obsessive. Although they may become relatively fixated on certain things, they can bring reflection so that innovations are thoroughly explored (philosophers even find a place in the mix).

For good and ill, we move beyond childish things, and not-so-childish fixations, and learn to see with broader vision and alternative perspectives. This truly allows us to enter into discourse with others and appreciate their insights. Thus, we learn from those who are oppressed and disadvantaged things that we have previously not appreciated and thereby we become reflective in new ways. We are sometimes humbled by the self-images that emerge and, in the process, become better people. This may be called 'repentance' by many faiths, a way-post on the road to personal salvation or enlightenment. It exposes us to 'the light of love' and the life experiences of others in an inclusive and dynamic triangulation.

Inhibition and Selective Attention

These neurocognitive tendencies may seem to potentially create a conflict between individual development and social straight-jacketing. If developed in a punitive and negative context, it can damage the affective and socio-political development of the individual and create a harmful context. A context of collegiality and mutual positive regard, however, transforms disagreement into healthy mutual critique, clarification, and enlightenment. A version of that is also applicable to children; they can learn to live (*disce vivere*—one of our old school's motto) in a loving and mutually inquiring context or one that differs from that in important ways. As parents, we must choose and therefore live wisely with these attitudes in our conscious or unconscious minds.

Karma ensures that 'the truth will out'. When love and truth mix and meld, the Aristotelian mean between inhibition or self-restraint and enthusiastic participation will be achieved spontaneously as it were. Such spontaneity tends to engender creativity that enhances our shared lives. We shall return to this theme when we consider human goodness as a natural disposition in the next chapter.

The art of judicious self-governance through a balance of inhibition and directed attention comes with maturity. As the old wisdom goes:

> When I was a child, my father was a God. In my youth I thought he was lovable but so behind the times and stuck in his ways. As an adult I was increasingly amazed at how much he had learned, almost as fast as I had. These days I listen a lot more.

This might apply even more and in a differently inflected way to mothers.

Our lives together are nested in emotionally laden exchanges and expressions. A smile can mean many things. Often it merely makes a link between us. Yet a wry smile can indicate that one has become jaded. Throughout life, we progressively learn, like young academics, that "to everything there is a season" so this prompts many 'turns': perhaps towards the skilful and well-judged cognitive flexibility of a mind that is receptive to and able to be selectively critical of ideas in the light of history. Like a well-practiced and apprenticed response, this scholarly critique brings rewards that are cumulative. It allows us to move forward according to *festino lente,* as a more measured pace has greater rewards in terms of fertile and evocative ideas. To develop and then reflectively explore this treasure trove, we need to be both respectful and inclusive, examining ideas for their value and the insights—albeit partial on occasion—that they bring.

We should therefore return to Kristeva and her verdict on complex and structured theories of language. Kristeva reflects on the work of Noam Chomsky (1977) who, for those recoiling from the austere causalism of Skinner and the behaviourists, was a 'ray of hope'. She remarks:

> Transformational grammar is carrying out in a more marked and revealing fashion the same reduction that structural linguistics, and especially American linguistics, effects in the study of language. Pure signifier without a signified, grammar without semantics, *indices* instead of *signs,* the orientation is clear-cut, and even more marked in Chomsky's unpublished works. (1989, 260)

This attempt to excise the speaking and engaged subject from the anatomy of natural language is one of the dangers of neurocognitive theory. She goes on "the very concepts of 'subject', 'truth', and 'meaning' will be dismissed as incapable of resolving the order of indicative language"(261). Thus, in doing properly naturalistic human neurophilosophy, we must reflectively 'get down and dirty' in the real world.

Consciousness and Cognitive Focus

'Consciousness' and 'cognition' are two terms that reveal the philosophical contours of our general theme of the soul. Those are revealed in two ways: neurally and conceptually.

Neurally in terms of synapses and cells; anatomically in terms of brain structures and regions; physiologically in terms of brain rhythms and circuits and somatically in terms of the neural representation of body regions.

Conceptually in terms of holistic relations between ideas and modes of cognition, relations based in the history of thought or scientific disciplines, styles of philosophy such as Anglo-American, Continental, Indigenous (of various lineages and cultures), artistic, etc..., etc....

In each of these modes of cognition, there are developing, mature, and fully developed varieties, and sometimes a person needs to take a step back in order to learn and progress. Many sources of wisdom have a phase of 'negative consciousness' or 'becoming like a child' or even 'being born again' as a first step in the path of overcoming developed prejudices and habits of the heart. Some of us are so arrogant and full of '*chutzpah*' that it takes a stroke or a stubborn animal companion—'Balaam's ass' for example—to get a lesson through and rescue us from folly and its praise.

Kristeva again:

> As a signifying system in which the subject *makes* and *unmakes* himself, (sic) language is at the centre of psychological and more particularly psychoanalytic studies. (1989, p. 265)

She concludes by suggesting that the study of language must "confront its limits" so as to "provide a more complete vision of linguistic functioning" and "take into account the subject, the diversity of modes of signification, and the historical transformation of these modes in order to re-found itself in a *general theory of* signification" (32).

Throughout life, we are fashioned by words and those who share them with us and immerse us in them. Childhood influences, as one might reflectively realise, are profoundly influential and less reflectively moderated in their influence and cognitive content. In later life, we might contemplate them anew and therefore moderate the effect they have had on us. Thereby we are changed and, one hopes, improved, but sadly sometimes we have been wounded or disfigured.

Wealth and youth, sadly, sometimes enhance our folly and both revel in the company of associated traits and states such as wantonness, forgetfulness, ignorance, and inebriation (Erasmus). If we are well brought up (or sometimes if we 'just grow'd') we can learn from the school of life and good colleagues, then we see the folly of these youthful things and eschew or moderate them. But that is a topic for our concluding chapter on goodness.

What our children learn depends on their social context as much for its affective and relational tone as for its content. If the children of the rich appear to thrive while those of the poor are stunted, we are all worse off and our shared discontents trouble us, as Freud and many moral theorists have noted. But this development, in human terms, combines cognitive and affective learning in intertwined ways. We can learn in a self-centred purely cognitive way, or more deeply and relationally. The latter makes us more inclined to identify strengths than weaknesses in those we encounter and to seek synthesis rather than opposition in acquiring knowledge. Here, as elsewhere, 'the medium is the message' and analytic philosophy often encourages destructive criticism, seeking thereby to emulate Socrates in one respect while at the same time neglecting his pervasive thirst for learning by discovering his own errors and perception of overly quick generalisations. Rather than 'constructive (or often destructive) critique'; which, can 'throw the baby out with the bathwater' one can be mindful of the Aristotelian mean—healthy evaluation, sensitivity to broad-brush generality, and interpretation within a generous frame of mind.

We might express the mean as considered receptivity, combining critical appraisal with willingness to learn in a way which develops the healthy intellectual habit of transcending the shortcomings of opposing tendencies, each of which is a worthwhile perspective if cognitively absorbed in moderation. In science, we signal such moderate acceptability with disciplinary devices such as 'statistical significance' and 'confidence levels'. Yet, in everyday life our considered judgements must be hemmed about more memorably with expressions that resist the attitude 'near enough is good enough' while accepting 'don't let the excellent be the enemy of the good'. In surgery, these are apt. In discourse we imbibe such stances and somehow do the informal work of striking the mean in our intellectual lives by attaining practised and refined conceptualisation and moderated interrogative acceptance.

In growing in the understanding of life, whether as academics or as infants, we are either nurtured or repressed according to the ethos in which we learn. That lesson is well attested by traditions of varying provenance. It has been commended by many different commentators on the human condition, whether iconic, historical, or fabled: Plato, Jesus, Balder (the bright and beautiful), Moses, Mahatma Gandhi, Māori Marsden, and others who are less well-known in the general and academic literature.

Anticipating how others will act and a willingness to cooperate with them is based on trust in their decision-making. We are engaged in joint ways of being in which our efforts are multiplied and complemented by others so that our abilities are extended in many ways. Thereby we achieve both co-operation or doing things with others, and 'con-scio-ness' or knowing things with the aid of joint activity and communication with others. This holistic, open-ended, and shared mode of being-human-in-the-world is inclusive, temporally extended in life and history, and constantly evolving and changing. It involves '*psuche*-ness', not 'spookiness', or the mystification that follows from mystery-mongering in trying to match consciousness with sub-atomic events in neurones.

Mirror neurons, other neural structures, and non-biological factors ground a mentalising network that enables human cooperation and trust in social interaction and our dealings with each other (Albertini et al., 2021). They help constitute the social brain which transcends mechanism in favour of interwoven rhythms and a dynamism that structures our lives and, ultimately, our life-world—a lived context of being human (Gillett & Franz, 2016). Evolutionary neurology, responsive equilibrium, and the moral brain go together. That context and its embedded relationships ground our structured social and moral lives with an adaptive resonance, as in Adaptive Resonance Theory (ART). As noted in Chap. 2, this theory involves a mathematical model of brain function as a natural tendency to maintain a balance between the need to process new information and the need to hold stable representations of the world (Grossberg, 2013). ART aligns with the neurodynamic models of Freeman, Buzsaki, and other neuroscientists.

These networks and structures form complex circuits of practised response to contingency. They draw on our observations of other creatures (particularly those like us and those we domesticate as we ourselves are domesticated (Wilson, 1988). Our empathy with them, our understanding response to their communications, and our shared implicit complementarity in rhythms of adaptation (think of working dogs or, to evoke a vivid fiction, Rudyard Kipling's named animals—such as Baloo—in *The Jungle* books). This powerful and intertwined complex of abilities, perceptions, and emotions enables us to form collectives whose members both real and fictional complement each other in the daily routines of adaptation so that, as a group and as individuals, they develop.

Functioning well as a member of such an adaptive alliance involves a set of sensorimotor and cognitive abilities that we call 'goodness', a kind of excellence of human natural function (Foot, 2001). Goodness, construed in that way, is continuous with many competencies comprising our human way of going on. Thus, our best thinking can be rendered effective in real life as a way of coordinating the activity of a group of human beings. That is why concepts like *akrasia* (including both 'hot' and 'cold' varieties of weakness of will) help us evaluate our current ways of going about trying to live a good life, one in which we do those many and varied things that tend towards human flourishing.

Akrasia is manifest when the agent acts in a way that their balanced self would not (Aristotle, 1984, Bk. VII). They can be unbalanced because of emotion ('hot'), or calmly intentional but do what the best self to which a person might aspire would not do ('cold'). Either way we show weakness in not doing what, in our right mind, we would do. In this *phronesis,* or balanced practical wisdom, virtue brings its own rewards.

Our reward, we could say, for doing that which properly engages us with others is to be enabled to do the things which make each person feel that they are exercising their strengths and doing something for the good of the whole. The resulting action may feel hard, stern, or 'holier than thou' despite it being the right thing to do in all-things-considered human terms. In developing the associated character that we call 'strength of will' or 'integrity', the highest affiliations of the self can guide and rule our 'lower' inclinations. The shortfall that beckons may have redeeming features. An example is parental—or often grandparental—sympathy and love for children that can tend to indulgence.

THE INTENTIONAL ARC

In thinking about the fluidity and flexibility of human cognition, it can be helpful to consider what has been called 'the intentional arc' (Gillett & Seniuk, 2018). The concept refers to a dynamic and flexible connection between the thinking of the subject and the subject's action in an ethological context to which the subject is adapted. As it is discussed in analytic philosophy, intention refers to the cognitive or psychological framing of an act. Any action is said to be intentional or purposive if one can discern a conscious thought that guides the action to achieve a certain end or serve a certain purpose within the mental economy of the agent.

The intentional arc links what the agent actually does with the whole structure of psychological life which shapes the agent's cognisance of, adaptation to, and dynamic interaction with an ethological setting. The intentional arc is a loop from holistic consciousness as the basis of mental life and an active search for a properly realised context potentially integrating the action and the agent's interaction with the environment. It departs from the passive reception view of perceptual processes that trigger cognitive and motor states causing an act. Instead, it involves an active search for information already woven into our adaptive pattern of going on. In that sense, it is thoroughly existential because it distinguishes the perceiver from a passive receiver and turns perception into an active process of seeking for the conditions in which a practised way of being oneself is fitting.

That is the arc *from* active search in the world *to* action and potential action in the world that is related to thought and the framing of purposes. The link to evolution is thereby clear in that habits of search adapted to a particular environment develop genetically and epigenetically to pick up what is of vital importance for the survival of an animal, and ultimately, of a species in that environment. This is dramatically shown by Australian aboriginals who can survive and even thrive in an ecological setting that becomes adverse and then incompatible to life because of colonisers who do not 'go native'. It also reveals an unheard voice in the political debate initiated by Plato under the name of Thrasymachus in Book 1 of the *Republic* (Plato, 2004).

The debate concerns political power and influence and suggests that it should always favour the strongest who have proven themselves most fit to govern society. But perhaps in an era where the world is undergoing unprecedented change, a new voice should be heard. Just as the Australian aborigines can teach us a great deal about sustainability and living within a demanding environment—the Australian Outback, so the human race must learn to live in harmony with rather than exploit our ecological context.

Overcoming the challenges of the environment has previously been a successful human strategy in which our genetic and epigenetic endowments have fitted us to maintain the position of an apex species. But we are in a time when sensitivity to the living family of earth-dwelling creatures (including plants) is that which will serve us best. In that endeavour, the weakest among us have a great deal to teach us all. They must learn to live with an environment that can sustain them but does not easily do so rather than rely on their naked power to extract from it and exploit it in an unsustainable way. To learn their lesson, we must be exquisitely conscious of and sensitive to the delicate balances which allow human beings to flourish and avoid the imbalances which can distort and ultimately destroy their chances of living well.

Consciousness allows us a wide discursive elaboration of everyday being-in-the-world. It involves a kind of knowledge *in actu* in addition to a kind of knowledge *de facto*. The difference is explored by Merleau-Ponty, as he analysed a human being's knowledge of his/her own body and actions (1945/1962, pp. 102ff). Immersed in being and doing as one is, individual active and reflective life is different from observation of self or others as mere elements of an objective event in a context. One's actions perceptions, attitudes, and orientations are more than just documentable events in the chain of causes. They are lived and subjective makings of a world that springs partly from oneself through the enacted rhythms of one's own brain and nervous system. This is one's incarnation as a being-in-the-world, a being with considerable powers to affect the world. To this, we add being with others. This form of being is also provided by discourse in the many ways that we have already discussed.

As we contemplate the 'knowing with' (*con scio*) that only becomes possible when we genuinely and empathically interact with those who struggle, we learn a certain humility and a certain respect. The new learning departs from fitness as we have conceived of it to this point. In casting doubt on our traditional notions of fitness and emphasising compatibility with the fragile world and its limited parameters within which we must live, we are forced to re-evaluate our values. Accordingly, might is no longer right in either of the senses of that term. 'Might' is not an access term giving us 'the right' to do certain things of which we are capable and can conceive. Nor is it 'right' in the sense of maintaining human quality of life and opportunities to improve our being, merely to use might to merely satisfy desire or exercise the will to power. What is more the might involved is overwhelmingly seen as might over others and not self-government in the pursuit of human excellence.

Being encompasses perception or 'seeing as'. That involves cognition not only in sensation but also in the mental acts of imagining and creating and the physical acts that flow from them. As we have noted, our cognition is inflected by culture, society, and politics. Wittgenstein was aware that a narrow conception of thinking or cognition compatible with human artefacts and their limited and constrained existence serving artificial aims could not be a suitable conception with which to understand human ways of being. A wide scope of 'seeing as' is available through imaginative constructions of possible experience. We have highly articulate techniques of sharing these through grammar broadly construed to include language and stories or visions infusing the consequent cognitive structure that is shown by thought and behaviour.

Thought-infused experience has been reconstructed for neurophilosophy in terms of anticipations of perception, predictive processing, predictive coding, or 'prediction error minimisation' (Kilner et al., 2007; Friston, 2018; Bastos et al., 2020; Milledge et al., 2021). This contemporary work has revolutionised neurocognitive theory but, as we have argued, has older though unclear and apparently semi-idealistic philosophical roots. The deeply evidential scientific realism in brain dynamics and biological life-in-the world is satisfying for those who are sceptical of post-industrial scientific realism but exhilarating for followers of Hughlings Jackson. That exhilaration, as a Thomist might point out, is heightened by a unique form of wilderness experience and 'coming in from the cold' by re-inhabiting warm cognition that nurtures flesh and blood.

The philosophical idea of anticipation of perception was originally suggested by Kant and taken to betray his idealistic leanings. It receives salient contemporary neurophilosophical support. The experiential basis of cognition and thought becomes either:1. caused by or 2. loosened warm cognition from, experience in something possibly like the way Frege dimly conceptualised it through *a priori* logic. Cognition and perception are inked to lived and *heimlich* experience in a demonstrable but surprising way—'a pinch of salt' indeed. *Ex hypothesi* there are rhythms of holistic brain activity involving multiple centres and circuits. They are ready to be enlisted for encounters between the organism and the world it finds 'around here'.

The current approach is more ethological and naturalistic than many past approaches. It avoids idealism while acknowledging the creative role of the subject. Rather than cognitive structures *per se* as envisaged by Kant (1781/1963) and Frege, human beings come to neurocognitive interaction with the world equipped with ways of seeing and construing what they experience. These ways of seeing are jointly fashioned by our lives in an ethosphere and how we have configured ourselves in adapting to the environment. The rhythms involved in this autopoiesis (or self- making) affect all aspects of human neurocognitive lives. They include notional, imaginative, and 'Wonderland' aspects, as the prominent mathematician, logician, and Anglican deacon Charles Dodgson (Lewis Carroll) noted (Carroll, 1865/2015).

We make up stories and communicate them, thereby planting in each other's minds ways of going on to try out and then experience. This enculturation allows top-down formation of a self in response to the possibilities of which one is able to conceive or vicariously share according to the traditions of the human group to which one belongs. This is especially appreciated by indigenous societies. Tradition can be rich with holistic and historical inflexions that may be traditional or quite singular. Folkways convey memorable events as they are nested jointly in deeply individual habits of thought and action and skeins of tradition. Such 'skeins' carry the characteristic flavour of a culture or an established 'technology' (such as 'blade-making'). Thus, holistically equipped in complex ways, each of us creates our own distinctive way of being somebody, somewhere.

One's consciousness is therefore not confined to what one actually encounters in biological life. Rather, through discourse and the envisaging of future life-moves, it enables us to form intentions which extend, allow us to reflect upon and execute ways of acting. These not only exploit our extant neural rhythms of interaction with the world, or those which we have received and internalised. They also extend them. The extensions are creative, fluid, and woven together. We do not do well in post-colonial thought to dismiss indigenous and historical cultures for what we regard as 'mere storytelling' or myths and legends as these are repositories of cultural memory. They become part of the individual and collective psyche.

NATURE, CULTURE, AND SITUATED PRACTICE

Wittgenstein was a misfit in many ways and an exile as some have portrayed him (Klagge, 2011). He was an exile from the land of his birth (Vienna), an exile from his culture and adopted culture (Cambridge) and the faith it embodied. He was an exile from the scholarly and academic unworldliness of Oxbridge (in its eastern incarnation among the fens of Cambridgeshire) and an exile in the subtle Christianity he found there in the various forms of Anglicism. This shows in his philosophy. His thought is at once semi-mystical, elusive, and 'down to earth', or ethologically realistic. He departs from the elegant and mathematically 'realist' (in the metaphysical rather than 'common sense' form) rooted in Fregean idealism, logicism and its Russellian, almost scientistic, realisation, and instead

embraces the biological/anthropological 'messiness' that comes with eth-
nic and bio-psycho-social life and the rhythms incarnate in living as a neurally advanced highly evolved organism.

Functional and dysfunctional ways of being human therefore become
tied to our ways of going on both naturally and culturally. They are not
tied to any functional, quasi-scientific, or logico-mathematical abstraction
conceptualised as 'rule-following' read as a 'functional procedure' involving individuals or communities (Gillett, 1995). Rather, we inhabit situated
forms of life in an ecological setting so that, even were almost all human
beings to lose their colour vision and subsequently come to regard it as an
illusion, their colour-sighted practices and surviving practitioners with the
ongoing ethological advantages conferred by that ability would convince
sceptics. Perhaps all our apparently merely artistic spandrels should begin
with that status in our analyses.

One can read Wittgenstein as responding to the post-industrial England
and Europe of his youth with as many approvals, misgivings, and reserva-
tions as one might think of Merleau-Ponty as responding to French/con-
tinental idealism inspired by the neuroscience with which he was in
contact. Both are reactive, quasi-heroic, but deeply unsettling, paths
forged in the heat of academic discourse *en route* to an inclusive naturalistic synthesis.

Wittgenstein can be seen as arguing for situated practice guided by
culturally inflected cognition. Merleau-Ponty speaks of the lived or "affective body" as prior in human experience by contrast with the objective
body of grammatical reference or scientific discourse (1945/1962,
p. 110). In so doing, he returns us to the first things from which we derive
'data'—the abstract starting point of much science. We apprehend the
phenomenological motto "back to things in themselves" (the *Ding an
sich*). We forget that insight in philosophical or scientific empiricism and
therefore halt part way on the voyage toward those 'first things' in a way
that indigenous thinkers do not.

Ethos, ethics, and law can be rendered in discursive terms in such an
inclusive framework. Ethos is the way of being realised in a given culture
and best seen, as we have noted, in Indigenous cultures. The Māori of
New Zealand or the First Nations people of North America show many of
the characteristics that need to be reflected upon and perhaps learned
from. In one's *Mihi*, or self-introduction for Māori, one is supposed to
mention the water, the mountain, the family, tribe, and people, to the
many-parted name are all important and place one as a being in the human

world. This is just as a fjord and lineage might be in Scandinavia or Norse settled parts of the New World. These locate us not just in terms of bare geographical coordinates but in introducing ourselves to others as we indwell and spring from the landscape. This is a way of contextualising one's name that articulates with history, the path of one's journey to the encounter, and the story of humanity. In being human in this way, we are being *heimlich* or 'homely'.

Māori Marsden puts it thus:

> Primacy of value in descending order is accorded to the spiritual, psychological and biological.
>
> Spiritual values ... are ultimate and absolute in nature and yet always beckoning man onwards. The closer one approximates to the ideal the greater the satisfaction. There is always a gap between the ideal and practice; between becoming and being but towards that excellence all things strive. (2003, p. 39)

Respect for nature or the ecosphere and its encompassing sustaining reality is part of our affective and therefore holistic neurocognition and self-awareness. This we share and can come to embody when, as colonisers, we 'go native', finding in ourselves an appreciation for 'the ethos of this place'. It is not that the 'echoes of home' grow less but rather that they change, and we see them as we have not done before, even value them more. Perhaps we also see them more realistically, bitter-sweet (as it were) after 'dancing with wolves'.

A sense of belonging is part of that new state of mind and habitus. We are changed in a way that is too easy to caricature either in a saccharine or cynical portrayal depending on many things including our attitudes of heart and mind (among which are 'reactive attitudes' (Strawson, 1974), arising from our origins and based on neural dispositions inherited and produced by upbringing. The authors (from New Zealand and US/Canadian backgrounds) have both had to take the first nervous steps of learning to swim in these subtle cultural currents, often generously aided in that task by indigenous scholars.

INQUIRY AND DIFFERENT TYPES OF KNOWLEDGE

A sense of shared life resides in that generosity. It is also intrinsic to human neurocognition against a backdrop generated by a marked departure from egocentrism as a partial distortion or degenerate form of the neurophilosophy of affective life. We live, in reality, by the support, nurturing, and affection of others and only reductive partial amnesia displaces that by egoism. Of course, all ideologies are selective, and academia is often seduced by fashion. But Wittgenstein's respectful agnosticism should serve as a lesson here. Self-critique rather than partisan criticism is always in order, along with a suitable seasoning of generosity of spirit.

In this regard, Merleau-Ponty speaks of a poem or work of art as a living thing (1945/1962, p. 174), revealing to all a reality that is thoroughly human and distanced from oneself when rendered as an object of gaze or contemplation. Science then takes up the reflective refrain. In the resulting abstract and objective creation, its own artificial form as an intellectual artefact, we submit the body and its sensing and movement to the scientific gaze. We then embark upon a study of human beings in the form of physiology or anatomy or biochemistry (or one of the other '-ologies').

A particularly powerful '-ology' is theology and its relatives. These include mythologies and cosmologies of various totalising kinds, against which thinkers like Nietzsche, Māori Marsden, and Michel Foucault warned us. When we consider the human condition, replete with stories, we look beyond the academic profile of these figures. We look beyond these profiles to explore the human life they partially disclose. Take for instance Nietzsche. He was the 'stranger in a strange land' if ever there was one, though steeped in a culture that he negotiated with anger, impatience, and penetrating insight. Even as his culture explored the natural origins of humankind with a post-enlightenment optimism, he inhabited the world of irony as indicated by the title *Human, All Too Human: A Book for Free Spirits* (Nietzsche, 1878/1996). It is ironic that Nietzsche himself was probably experiencing the initial effects of a brain tumour as he wrote the work. In it, he echoes themes typical of him: the failure of Darwinism and geneticism, the importance of the revolutionary and the individual thinker, the free soul, the deviant, the artist as cultural product and cultural misfit, the saltatory and twisted strands of human excellence in history, and the false seduction of what we call 'progress'. These themes disrupt the European civilized dream, so aspirational in the North America of Henry James and his brother, William. So, from that tradition we can turn to one encountered more recently in history but reaching back unbroken into prehistory.

Indigenous thought and its revival in the Antipodes has begun to join the waterfall of ideas of which Nietzsche speaks. Māori Marsden, whom we have briefly mentioned, speaks like an Old Testament prophet to the First World aspirational philosophy and theology of New Zealand and the colonies. His words ring with themes resonant in Māori minds despite the unfamiliar Anglicised and Italianate vestments in which they appeared in the Antipodean setting. Māori values underpin and are embodied forms of Western abstractions. They recognise the divine power in *ihi,* the living authority *of mana,* the propriety of awe—*wehi,* the sacred—*tapu,* and the need for rites—*toha*—to honour these things. All these are part of Māori— the life principle—much older than Western science or even its Greek origins. This is not a setting merely for survival but quintessentially for *aroha*—love—which issues in four (or maybe more) 'loves'. That branching reactive attitude—of emotions such as empathy, compassion, and forgiveness—is a human reality with a genealogy and an associated set of *praxes,* is the basis of morality.

Foucault might be thought to intuit this resonance in part and without the indigenous link. In one of his later works, he reflects on the intertwined nature of control of the self and the demands of society expressed in social norms and socio-political structures. He unfortunately, but understandably, focuses on 'two loves' not the 'four' of C.S. Lewis (Lewis, 1960). However, the earlier (and Christian) Lewis does recognise reciprocity and commitment even when *eros* is the focus. These constancies endure and bind us together, perhaps as an early adaptation to offset one source of destructive intragroup conflict. Whatever the origin, there is no reason why the human neurocognitive system might not elevate such powerful blends of cognition, emotion, and socio-political artifice to forge those practices embodying fellowship and collegiality. This can all be part and parcel of the social brain. These creations are interrogated in Foucault's essays on ethics (Foucault, 1997).

In his discussion of sexuality and solitude in that work Foucault remarks as follows:

> I have tried to get out from the philosophy of the subject, through a genealogy of the modern subject as a historical and cultural reality—which means something that can eventually change. (1997, p. 177)

This theme of the cultural-historical context supplements the thoughts we have drawn from Kristeva to embrace the dynamic historical ethology of the subject. This is not a posit or a dimensionless point of subjectivity but a mortal embodied speaker and thinker sharing life with others in a fragile world which we, individually and collectively, inhabit and in which we have significant power. That power is frightening and perhaps suicidal and genocidal. In our time, this action in the world meets us in the guise (*inter alia*) of climate change, in relation to which the destructive impetus of human life is generated by shared socio-political power wedded to self-interests of all-too-evident, if not obvious, and privileged kinds of human beings associated with 'astigmatisms' or 'cloud cuckoo land' optimism. These form a 'demonic' alliance that is so pervasive among us that its reversal is almost unthinkable and is urgently becoming impractical.

This is our contemporary 'Loki' or 'mountaintop' moment (Matthew 4:9). In the biblical story, Jesus is tempted by Satan with power over all the nations of the earth if he will only be political and worship the embodiment of human corruption. Think of the good that could be done with such power! If we do accept complicity in an evil path for us all, what a wonderful mental or moral web we are tearing apart that we might otherwise weave. Perhaps we should eschew the thought the occasional 'head would roll' if the motive and vision is good and just—as in the 'reign of terror'. We could instead, against all reason, summon the resolve to think the unthinkable, remake ourselves and society so as not to fully realise the folly which is our modern 'Babel' moment (Genesis 11).

A sense of mortality sobers us when we think of the intertwining of love and death. The tragedy of death, its shared reverberations, and the fond memories it occasions, render human life mortal and fragile unlike fabled Olympian power and invulnerability, as AIDS has made evident (Gillett, 1987). We have met this scourge of us all together and learned deep ethico-political lessons about human diversity, marginalisation, and privilege. Together, we have begun to turn back a tide that power and authority in the socio-political world cannot alter by *fiat*, though that was famously tried in legend.

Our modern 'Babel' moment (Genesis 11) has similar psycho-social roots to the depicted setting. We speak a common language—that of commerce. We recognise common self-advancing tendencies. Some of us conspire to support political hegemony, despite corruption. We exploit the earth's natural resources and see that as a way to universalise well-being. And we idolise abstract goals that 'reach for the sky'.

Some of this is keenly realised in the Norse fable 'Balder the bright and beautiful' and in the Easter rites. In each, human imagination encapsulates a sobering lesson. 'There is a kind of 'evil' which cannot be changed, and its undoing is unthinkable or 'a miracle'. Neurocognition constantly strives to give the miraculous a thinkable form. But perhaps instead we should look to a knowing deep within us that is not in a logico-mathematical and scientific uniform: Rangi, Papa, and their offspring; God, The Holy Virgin, and the incarnate Son; Almighty Father, Incarnate Son, and Holy Spirit. These images have power. That power predates human emergence as world-changing and infuses that ability with humility. We do our best to note the still small voice within that says to us "What are you doing here?" as we hide in our philosophical cave from an apparently overpowering empire of the mind and spirit formed by reason emerge.

Prior to and prescient in relation to Foucault's socio-political turn, Merleau-Ponty also begins his philosophical dissection of the ego with the science that is its setting (as does Kristeva) and both move on to social being and the language that conceptually locates the ego. We must recall, however, that social being and the cognitive precision or abstraction afforded by language take place amidst the ongoing and dynamic "intercourse between the body and the world" (Merleau-Ponty, 1945/1962, p. 78) where we achieve a *heimlich* state that is 'being in the world' as a creature who is not only both sensorimotor and affective but also both cognitive and socio-political. To these attributes language and imaginative thought add extension in time and space beyond the actually experienced so that each of us becomes a creature of wonder and imagination. This expanded being we manifest to 'the other' throughout life and beyond it, if only to readers with whom our mortal lives do not overlap. That wonder is rendered precious by the fragility of being mortal and concerned compassion emerging from the entwinement of lives whereby we glimpse each other's vulnerabilities and affections. This is the sensitive core of love (*aroha*) and this, we suggest, is the heart of ethics and religion.

Religion always begins with the mystery of origin. From that primeval beginning emerges the mysterious power and manifest authority of the father. It is imbued with the role, if not the sexuality, of Freud's figure, a version of that which appears in many mythologies. This figure, however, is prominent against the mystery and the undeniable life-giving/supporting/sustaining presence of the mother—'mother earth', *papatuanuku* (Māori), the ground of our being. There in the beginning, and unto our ends, mortal or otherwise, we find the maternal witness and loving support of life both male and female.

Theology also can find a place for such things, though perhaps not one that is as doctrinal or as dogmatic as post-enlightenment neurocognition would suggest. So, we are reminded again of the fluidity of Merleau-Ponty. After touching on sexuality and the lived body, he comes to the cogito and our cognitive lives together. His approach is both salutary and seductive as it rings true to twenty-first century neuroscience and dynamic neuroimaging rather than the more limited version that is often selectively cited to support mechanistic models of twentieth-century neuroscience.

Existentialism is the emergence or 'standing out' of the young man par excellence (Jaspers, 1938/1971; Sartre, 1945/2007). He makes his own way in the world by self-definition, not only as 'I am' cognition but in addition as 'I emerge' perhaps supplemented by 'and I do so' 'leaving childish things behind', though sometimes needing to be reminded of the precious and vulnerable things of childhood. To rediscover ourselves as children is to moderate our emergence as adult and responsible beings who have a role in the world. Part of recovery is to become openly responsive to and appreciative of others *sans* classification. We re-cognise (and regret some features of) ourselves anew. This is not merely 'reactive to' others in the mode of formed cognitive 'slots' or practiced neural rhythms but involves empathising with 'the fallen', even finding ourselves among them.

If theology is to be consistent with neurodynamics, then it would be strangely related conceptually to a topic addressed by Wittgenstein in his lectures on aesthetics (1967). He begins his remarks reflecting on the word 'beautiful', which he notes that, although it has the form of an adjective and thus seems perhaps part of completing a specification of the evident appearance or describing a quality of an object, it expresses an evaluative judgement like 'good'. He ponders the uses of such words and the manner of learning them. Both are embedded in a network of experience, evaluations, and cultural encounters, all of which are historically embedded and socially inflected. This is not a mechanistically causal phenomenon but instead pertains to 'the soul' (p. 29). Our reactions to each other and to pictures are replete with this complex of attitudes, which are reactive but only disparaged if we assimilate them to 'feelings or emotions' (p. 35). What, for instance is a kindly smile and what tiny change transforms it into an ironic one? How one sees the world when Paradise is lost actually teaches us a great deal. But the lesson is hard to describe.

The thoughts concerned here are subtle and deeply human. Similar subtlety should attend our attitudes to religion. The truth concerned is not authoritative in the cognitive sense and propositional or doctrinal like a quasi-scientific thesis. Rather, it has the authority of life-affirmation. It is not so much an '-ology' (like theology or ontology) as an underpinning framework for a reactive stance which is capable of confounding cognition with apparent contradictions and implicit reproofs. These are levelled at imperious human cognitive packages and generating paradoxes and contradictions, somewhat like Delphic oracles.

The burning bush, for instance, 'burns' but is not thereby consumed. One might ask whether our passions are sustainable or whether we are 'consumed'? (Exodus 3.1–22). 'Let him without sin cast the first stone' is applied to the fate of a prostitute. The men concerned in the condemnation are charged to be sincere and examine themselves and, as a result, none are left to condemn (John 8 1–11).The still small voice or 'gentle whisper' (1st Kings 18:20–40; 19:12) may convey a truth that is profound but cannot be shouted or proclaimed like a clarion opinion. Also, Jepthah's daughter warns us against grand human oaths and bargains. Beware lest our grandiose promises 'holding the world to ransom' bring us to deep and tragic folly. (Judges 11). A certain cast of mind could condemn even a loved child as a result of a pattern of cruel contingencies.

In each of these passages, a solemn human cognitive determination is unseated, transcended but not refuted by a more astute argument and not downgraded but genuinely transcended in service of a fuller conception of the good. Indeed, these problematic passages could provoke at least two responses:

1. One could suspend all belief and retreat to grounded and austere scientific thought. This route tends to overturn a long and many-stranded or multicultural human tradition in favour of a particularly propositional type of cognitive construction. It reaches its 'Waterloo' in 'QBism'.
2. One could preserve the inherent normative judgement for 'the spirit of the thing' rather than an overly cognitive or propositional construal and be wary of religious doctrines. There might even be an argument to suggest that an overly definite cognitively prescribed form of religion might lead to division and intolerance and that indeed there was an historical case to be made that it did so. One could even predict in a scientific way that it would lead to what

could be seen in retrospect as great and repeated evil in the form of religious war and persecution.

The argument can be syllogised:

1. Religions have both affective/relational content and cognitive content.
2. Human error can lead to over-valuing either.
3. Over-emphasising the first can lead to great division and interpersonal harm.
4. Over-emphasising the second can lead to a slackness of valuing human life.
5. The two must be astutely combined so as to avoid both faults.

Such is the case in marriage—ideally a committed partnership which forms a stable basis for child-rearing with both male and female qualities embedded in it. That ideal creates a positive life context, even if childrearing is not involved. The partnerships without the responsibility of childrearing can occupy all sorts of roles in the socio-political and cultural worlds we inhabit. Thus, the fostering *ethos* is valuable. Given that all our partnerships are dynamic, sometimes fragile, and variably tested in the eddies and currents of human social and relationship-based life, they need to be supportive of frail human beings. They must create warm and enduring contexts (the metaphor can be seen to be apt when one considered the wellbeing of mammals). Indeed, some cultures expect grandparents to take a key role in the stability of families and groups. That makes sense at a number of levels as time and tide have often worked away at sharp edges in their lives and cast a retrospective and reflective light on what once seemed so pressing and stressing and fraught.

Human society has what Merleau-Ponty recognises as personal being in a culture (1945/1962, p. 405). Such personhood in a context comes with a personality that is ready to be formed in all its neurocognitive complexity. That complexity, emergent from being and cognisant of thus aware of 'knowing with' (*con scio*) others, ushers in a kind of freedom to be which is dynamic and goes beyond the present, limited awareness of that which sustains it in a holistic way.

At this point, we can reflect on the appeal of dualism. We can usefully revisit a recollection by Bouwsma of an exchange with Wittgenstein about Descartes' 'Cogito.' Wittgenstein made a comparison with the cinema and reflected on the immediate picture of the moment likening it to the present moment of consciousness: "One could just gape. This!" (Bouwsma, 1986, p. 13), "*Thought* can be about what is *not* the case".

Here in the *Philosophical Investigations,* Wittgenstein inserts into his reflections on the picture theory of meaning. He uses it to inflect his ethological account with the fact that 'pictures' are creations that may not be photographic but resonate widely within human language and thought. Logic is part of this creativity and bounded only by logical possibility. This is well known to Charles Dodgson, among others. It is not that we 'cut off our heads' in this enterprise; we merely abandon sober ethologically situated reflection on real experience for an abstraction. Wittgenstein identified this *Tractatus* type of world and the impasses it gave rise to and turned away from it to 'get real' in the *Investigations.* In so doing he betrayed the philosophical purity of unrestrained logic for the situated real-world cognition aspired to by science (anthropology included). However, that turn in most hands abandons reflection on self-images of the age. These include all the causal machines, quantum realities, and mysteries, with a shifting line between them. They become enmeshed in themselves—their own constructions, interrogations, and inspirations about existence cognised and sometimes immune to the searching insights of existentialism.

'He would cut off his nose to spite his face'. On first encounter, this remark might seem opaque, but a single human interaction might bring it to the fore in one's mind. Thereafter it becomes iconic, capturing something about a person that no other expression does so well or so memorably; human life is full of such 'flashes' of illumination.

Conclusion

The inspiration of 'becoming yourself again' replete with culture, personal style, spirituality, 'and all that jazz' is comprehensible, especially through Merleau-Ponty. But it was prefigured by Hughlings Jackson. In his thinking and particularly his theory of recovery (York & Steinberg, 1995), Jackson stressed that, in addition to localisation in the brain, there were multiple levels at which any neurocognitive function was variously elaborated in the neural network. There were tract-related subcortical centres with limited sensorimotor roles. Neural connectivity ruled, however, with higher levels sketching in the details at specialist cortical sites and multi-level elaboration. In this way, the whole brain could be reconstructed after a local lesion by strengthening and building on surviving associative connections.

This account explains much including the multiply overlapping realisations of brain activity serving different functions. Under the influence of localisation models, these functions involve massive redundancies. The pared-down picture expressed itself in subtractional functional imaging and mapping of neural architecture. That enterprise served well in the process of understanding localised brain damage and thus aided clinical diagnosis and directed neurosurgery. But it did not illuminate recovery until Hughlings Jackson's corrective and holistic layering was also appreciated in the clinic. The clinics were not always as well-connected as the evolved brain. Indeed, clinical neuroscientists, including one of the authors, do not always realise what a cause for hope his prescient words are. He has learnt much from the first-person experience of recovery.

In order to move away from a path of increasing abstraction, we have tried to bring a lived-in neurocognitive grounding to this inquiry in order to ask what is good for human beings with an eye to the longitudinal narrative reality of our situated embodiment. We examine this in Chap. 7 and the human insights that enliven and illumine our conception of 'the good'.

REFERENCES

Albertini, D., Lanzilotto, M., Maranesi, M., et al. (2021). Largely shared neural codes for biological and nonbiological observed movements but not for executed actions in monkey premotor cortex. *Journal of Neurophysiology, 126*, 906–912.

Aristotle. (1984). *The complete works of Aristotle*, volume II (Trans. and ed. J. Barnes). Princeton University Press.

Bastos, A., Lundqvist, M., Waite, A., et al. (2020). Layer and rhythm specificity for predictive routing. *Proceedings of the National Academy of Sciences USA, 117*, 31459–31469.

Bouwsma, O. K. (1986). *Wittgenstein: Conversations: 1949–1951* (Ed. J. Craft and R. Hustwit). Hackett.

Carroll, L. (1865/2015). *Alice's adventures in Wonderland*. Penguin.

Chomsky, N. (1977). *On language: Chomsky's classic works: Language and responsibility and reflections on language*. New Press.

Foot, P. (2001). *Natural goodness*. Oxford University Press.

Foucault, M. (1997). *Ethics, subjectivity, and truth* (Trans. R. Hurley, and Ed. P. Rabinow). Penguin.

Friston, K. (2018). Does predictive coding have a future? *Nature Neuroscience, 21,* 1019–1021.

Gillett, G. (1987). AIDS and confidentiality. *Journal of Applied Philosophy, 4,* 15–20.

Gillett, G. (1992). *Representation, meaning and thought.* Clarendon Press.

Gillett, G. (1995). Humpty Dumpty and the night of the Triffids: Individualism and rule-following. *Synthese, 105,* 191–206.

Gillett, G. (2001). Signification and the unconscious. *Philosophical Psychology, 14,* 477–498.

Gillett, G. (2015). Culture, truth and science after Lacan. *Journal of Bioethical Inquiry, 12,* 633–644.

Gillett, G., & Franz, E. (2016). Neuroscience, psychiatric disorder, and the intentional arc. *Consciousness and Cognition, 45,* 245–250.

Gillett, G., & Glannon, W. (2020). The neurodynamics of free will. *Mind and Matter, 18,* 159–173.

Gillett, G., & Seniuk, P. (2018). Neuroscience, psychiatric disorder, and the intentional arc. In G. Stanghellini, M. Broome, A. Fernandez, et al. (Eds.), *The Oxford handbook of phenomenological psychiatry.* Oxford University Press.

Jaspers, K. (1938/1971). *Philosophy of existence* (R. Grabau, Trans.). University of Pennsylvania Press.

Kant, I. (1781/1963). *Critique of pure reason* (Trans. N. K. Smith). Macmillan.

Kilner, J., Friston, K., & Frith, C. (2007). Predictive coding: An account of the mirror neuron system. *Cognitive Processing, 8,* 159–166.

Klagge, J. (2011). *Wittgenstein in exile.* MIT Press.

Kristeva, J. (1989). *Black Sun: Depression and Melancholia.* Columbia University Press.

Lacan, J. (1977). *Ecrits.* Norton.

Lewis, C. S. (1960). *The four loves.* Harcourt Brace.

Macfarlane, A., & Macfarlane, S. (2019). Listen to culture: Māori scholars' pleas to researchers. *Journal of the Royal Society of New Zealand, 49,* 1–10.

Macfarlane, A., Macfarlane, S., & Gillon, G. (2015). *He Awa Whiria:* A braided rivers approach. In A. Macfarlane, S. Macfarlane, & M. Webber (Eds.), *Social and cultural realities: Exploring new horizons* (pp. 52–67). Canterbury University Press.

Malcolm, N. (1965). *Ludwig Wittgenstein: A memoir.* Oxford University Press.

Marsden, M. (2003). *The woven universe: Selected writings of Rev. Māori Marsden* (T. A. C. Royal, Ed.). Estate of Māori Marsden.

McConnell, D., & Gillett, G. (2005). Lacan, science, and determinism. *Philosophy, Psychiatry, and Psychology, 12,* 63–75.

Merleau-Ponty, M. (1945/1962). *Phenomenology of perception* (C. Smith, Trans.). Routledge.

Milledge, B., Seth, A., & Buckley, C. (2021). Predictive coding: A theoretical and experimental review. https://arxiv.org/abs/2107.12979v.1

Nietzsche, F. (1878/1996). *Human, all too human: A book for free spirits* (2nd ed., Trans. and Ed. R. J. Hollingdale). Cambridge University Press.

Plato. (2004). *Republic* (Trans. C. D. C. Reeve). Hackett.

Read, R. (2001). What does 'signify' signify? A response to Gillett. *Philosophical Psychology, 14,* 499–514.

Sartre, J.-P. (1945/2007). *Existentialism is a humanism* (C. Macomber, Trans.). Yale University Press.

Strawson, P. F. (1974). *Freedom and resentment and other essays.* Methuen.

Wilson, P. (1988). *The domestication of the human species.* Yale University Press.

Wittgenstein, L. (1967). *Lectures and conversations on aesthetics, psychology and religious belief* (Ed. C. Barrett). Blackwell.

York, G., & Steinberg, D. (1995). Hughlings Jackson's theory of recovery. *Neurology, 45,* 834–838.

CHAPTER 7

Being Good

Contemporary normative and healthcare ethics are both based on an abstract post–industrial method focused on technologies of production and associated self-interested behaviours. This framework of ethics gives pre-eminent place to autonomy or self-rule along with abstract or quasi-legal duties. It has generated rule- or principle-based reasoning as a complete model of moral behaviour. Our earlier remarks about operant conditioning suggest, however, that we and other creatures seek the stimulus conditions in which a well-practiced response to the world can develop or be expressed. This accords with the wisdom coming from a reflective life with a mind free of fashionable presuppositions such as the idea of an external reward for our efforts in aspiring to different goals. We need also to acknowledge our rich biopsychosocial nature in our place of existence and relationships broadly construed in terms of human ethology.

The behaviour-reward-punishment trilogy is best suited to law and abstract logico-mathematical thought and reasoning. It has proven much too simplistic and materialistic in a world where indigenous 'criminality' and self-destructive behaviour are widespread. These and other maladaptive behaviours and their ill-grounded conceptual links have intruded into philosophical psychology.

G. Gillett, W. Glannon, *The Neurodynamic Soul*, New Directions in
Philosophy and Cognitive Science,
https://doi.org/10.1007/978-3-031-44951-2_7

A syllogism seems in order:

1. Reward and punishment psychology is evolved and adaptive.
2. Human beings operating in this way do maladaptive things (crime and psychologically harmful mutual exploitation).
3. Reward and punishment thinking is a misleading abstraction from human adaptation.

This lays the groundwork for going beyond an operant conditioning-based emphasis on frequently emitted or species-probable responses and individualism as the true route to an adequate conception of the adaptive behavioural repertoire. A number of distinctly human goods then come into view: fellowship ("Oh how good and pleasant it is, when brothers dwell in unity"); love of your fellow men (and women)—("This is the sum of the law and commandments—love one another as I have loved you"); giving to others ("Greater love hath no man than this, that he lay down his life for his friends"). Throughout history, such things have inspired us and dwelt in our most cherished memories (as any war memorial testifies). Their occurrence in a 'sacred text' speaks volumes. We are quintessentially a species for whom the greatest good is self-giving for others. That is all the more remarkable as we are neurocognitively equipped to anticipate the future and envisage things only anticipated, whether for good or ill. This is truly a move beyond the animal, immediate and instinctive to something with more affective resonance, albeit counterfactual and aspirational.

In the post-colonial world, a particular type of individualism and self-focus has become a norm with a quasi-scientific hybrid naturalistic-normative status. This status, focusing as it does on individual material benefit, is a distortion captured by the above syllogism and the empirically informed discussion of it. We must be properly and openly scientific about our neurophilosophy and not be swayed by the enlightenment naturalism that tends to paint us as being 'red in tooth and claw' like apex predators at their most predatory. This may be a good reason to listen to women philosophers discussing a range of feminist views about the relation between the individual or group and the world (Lindemann, 2019; Maitra & McWeeny, 2022). That openness may be wise even though the females of apex species who live in family-based groupings often do much of the hunting work.

NORMS AS FORMS OF LIFE

Thinking in the post-colonial and abstract way, ethical guidance is already quasi-legal and propositional in its nature. This was originally based on Anglo-American codes of ethics. However, much of behavioural psychology, and the science of behaviour based on it, rests on the conceptual separation between an act and its reward. In fact, as we have noted, animals will work so as to be able to enter the conditions in which they have developed a highly frequent way of behaving. This result seems paradoxical to belief-desire psychology in which cognition is aimed at getting a material reward for an act or sequence of actions. It is not at all paradoxical to ethological thinking. Animals adopt ways of going on that sustain their being as individuals and as a species in a given adaptive milieu, however inhospitable that seems. Here the Australian aboriginal people with their ways of being reaching back beyond recorded time come to mind anew.

When one sees the opportunity to go on in a certain familiar way that is part of one's being as, in itself, the proximate goal of behaviour and as part of a mode of adaptation built on one's natural propensities, the enterprise of human life takes on a different character. As a natural species, we aim to do what it feels good to do (being 'in the groove'), and in that our shared language and discursive habits figure prominently. This language and these habits of interactive agency are forms of shared adaptation deeply ingrained in indigenous cultures. These are forms of life. They include Native American cultures (Arola, 2011), Māori (Stewart, 2021) and many others. The ideas of embeddedness in the physical environment and the relationship between the mind and the social world are much stronger in these cultures than they are in colonial cultures. Indeed, there is no strict distinction between individual and world in their notion of embedded ethological flourishing. However, shared adaptation through interactive agency applies to all of us. Ways of going on are part of the holistic profile which characterises each one of us as a human being. The goods of the soul may have a highly individual profile adapted to the particular embodied and embedded being who is becoming a person in relation to a range of interacting cultural mores with their diverse and variably shared socio-cultural orientations.

For instance, for a person with an excess of *chutzpah* (or self-assurance), a modicum of self-doubt might be in order, whereas for a normal person this self-doubt might be an impediment to achievement. What counts as a good human being can also undergo socio-political transformations in historical time. Thus, physical strengths and a bellicose nature is no longer considered 'good' in a human being, however useful they were when we competed with each other for ethological success. The benefits of extending intra-tribal tendencies to match the spread of our communicating contacts surely means that, as an apex species, we can resolve disagreements in a mutually constructive rather than destructive way. Indeed, at the present moment in history we have such efficient and indiscriminate ways of injuring and killing each other that we seem to face a forced choice between harmony and extinction. Our intercourse is so pervasive and frequent that this inclusivism would seem a vital contemporary human adaptation.

In this task, we have the great advantage of being able to form and reflect upon intentions to act. These can be articulated in such a way as their proximate and distal, in time and space, implications can be considered. We are sometimes prone to over-thinking and silencing those voices of intuition which carry the traces of past and collective experience as cultural memory and tradition. Indigenous societies do not let this happen to the point. Yet they sometimes risk being encumbered by the past. As a promising primate, this is when the benefits of philosophy in general and moral philosophy in particular can be clearly appreciated. Thus, modern highly structured reasoning is balanced, in law, by precedent as is seen in jurisprudence, a developed institution and *modus operandi* of legal reasoning. This provision relates a present case to past decisions in some respect relevant to it and which nestle reason about that case in a context more inclusive than the self-images and strategies of the present act. Our intentions become reflective and illuminated by a wider context.

Some of the intentions we form and express in discourse of various kinds relate us fruitfully to other beings within the context of our shared adaptation to the world. Within these shared folkways, we have various ways to be good human beings who thrive individually and collectively. We do this, for instance, by binding ourselves to each other through promises, jointly formulated strategies influencing what will happen so that present cognition potentially enhances our shared human life in the future, via the unspoken agreements that have been built into us through our development among others. These acts and undertakings curb our individual excesses and blind spots which would otherwise threaten the good. We realise that the good does not completely begin or even end with ourselves, however prone we are to that illusion. For example, we inhabit an ecosphere in which there is a balance that we must learn to recognise and respect. On occasion, that must curb what might otherwise seem like a good idea, like the use of the stored energy in fossil fuel. As Aristotle noted, the classical ideal of moderation and balance implies that virtues always involve propriety within the scheme of things. In contemporary terms, environmental sustainability is one form of this situated propriety.

MORAL NATURALISM

Philippa Foot explores a naturalistic account of ethics rooted in a conception of human flourishing that, as we have noted, is deeply Aristotelian (2001). Foot focuses on the relation of virtue to happiness and the relation of these states to reason, desire, and action. Patterns of virtue in people's behaviour are among natural goods, and virtuous behaviour is a type of moral excellence. Natural goods are not manifestations of a Platonic Form of the Good that exists independently of us and can be grasped by pure reason. Rather, they are natural in the same sense that health is a natural good in ensuring bodily homeostasis and the physical and mental functions that enable us to achieve a certain level of well-being. Foot's conception of human flourishing takes its origin from an ecologically situated view of the flourishing of any species with certain inherited and genetically grounded propensities. She aims to relate "the good" of a species to "the characteristics and operations of plants and animals"(p. 25), such that flourishing can be natural and linked to the grammar of our talk/language about particular types of things. In our case, that concerns the nature of our own species and "the relation of any individual creature to the 'life form' of its own species" (p. 27). Here the relevance of Wittgenstein's writing on grammar and forms of life is apparent.

Foot appeals to the later Wittgenstein in discussing the language of morality generally and the idea of good in particular. For Foot, the 'logical grammar' of the evaluation of human behaviour brings words back "from their metaphysical to their everyday use" (p. 3). Happiness (in the broad sense of *eudaimonia*) has a logical grammar that is distinct from the grammar of excitement or elation. It is thus a mistake to define happiness or *eudaimonia* as an individual state of mind separated from one's place in the social world.

This is consistent with Aristotle's point in the *Politics* 1, 2, 1253a2, that each of us is by nature not only a thinking but also a political, or social, animal, a *zoon politikon* (1984, Volume II). Happiness is not just a positive emotion associated with pleasure but a state of flourishing in which we initiate and complete individual and collective goals. These include both personal projects and public goods such as healthcare and a clean environment. Knowing how to go on acting in accord with natural norms is a social phenomenon. This is how we make moral progress in recognising and responding to reasons for respecting the rights, needs, and interests of others in cooperating with them. The reasons from which we act are not merely self-interested but reflect what we owe to each other and how our actions can be either mutually beneficial of mutually harmful (Scanlon, 1998). Widespread inoculation with a safe and effective vaccine to reduce the health burden of the SARS-CoV-2 is an example of applications of principles of beneficence and nonmaleficence that all could reasonably accept, or at last not reasonably reject. Actions that could reduce the adverse effects of climate change on people's health is another example, as is universal access to basic primary medical care. These are collective action problems whose solution requires cooperation on local, regional, and global scales. The satisfaction resulting from solving them comes from knowing that we can have a life relatively free of premature disease and death and still enjoy the liberty necessary to flourish by undertaking and being engaged in different projects.

It is worth repeating, however, that goodness is not reducible to following principles and reasons. The natural goodness of persons manifests in intellectual and practical virtues. In the *Nicomachean Ethics*, I, 7, Aristotle claims that to understand the human good, we must identity the characteristic function of a human being. The good for humans is rational activity performed well, which is activity in accord with virtue. Virtue is not limited to specific dispositions such as courage, patience, or moderation, but the general state of knowing how to act (knowing how to go on). For Aristotle, human good or happiness consists in rational activity reflecting the agent's conception of what is worth doing and the point of doing it (Korsgaard, 2008).

Agents are not isolated individuals and do not act in a vacuum but within a social network. What is worth doing involves not just individual but also collective action based on our being in the world with other human agents and social virtues of trust and fairness that are necessary to cooperate for mutual benefit (O'Neill, 2013, 2018). This is an advance over a more basic reductive evolutionary theory of social behaviour. In accessing limited resources, organisms progress from competing for access to these resources to sharing their efforts by following successful rules of decision-making in behavioural interactions between members of the same or different species (Taborsky et al., 2021). A similar strategy should surely be rationally pursued by contemporary humanity.

There are many types of natural goods. Friendship is an obvious example. Cultivating and maintaining a friendship is a cognitive-emotional skill. But while this good is not rule-based, it may not be capable of promoting and realizing moral progress in improving human behaviour because not all humans are motivated by regard for others. Still, we need a lower normative common denominator for day-to-day decision-making. Knowing how to go on in social interaction improves adaptability and cooperation in securing and enjoying public goods. This involves sets of actions that make us all better off and improve our individual and collective well-being. Moral progress is the outcome of a certain type of moral reasoning. Although it is distinct from Aristotelian ethics in some respects, the idea of morality as behaviour guided by a contract among rational agents offers an equally important way of understanding the normative aspects of knowing how to go on (Ginsborg, 2020). Natural goodness must not be confused with moral progress. It is not identical to principles and reasons to guide other-regarding actions. Nevertheless, moral progress is an implicit prerequisite for natural goodness. Our relationships with others and the world are not always innately harmonious as groups often begin alienated from each other ('Sharks' and 'Jets' spring to mind). We must be able to resolve conflicts between individuals and groups to attain and fully appreciate natural goods (sometimes through shared tragedy). Moral progress is necessary to distance ourselves from self-interest and self-satisfaction on the way to flourishing but can be painful.

The content of morality therefore consists in certain types of understanding, discourse, cooperation, and ideally agreement between and among human beings. The idea of agreement is the basis of social contract theory. It has two forms. According to *contractarianism*, rational self-interested agents cooperate with each other because it is mutually advantageous to do so (Hobbes, 1651/2017; Gauthier, 1986). The aim is to maximise one's own interests in a type of bargain with other like-minded individuals. According to *contractualism*, morality is grounded in the equal moral status and mutual respect of persons (Rousseau, 1761/1968). The moral status of persons is grounded in their capacity for autonomous rational agency, a view that is traceable to Kant (1785/1964). But contractualism is more than an advanced version of Kantian ethics. It does not involve categorical or hypothetical imperatives about how we should act in accord with the moral law, and consequences are not excluded from assessing the rightness or wrongness of actions (Foot, 1972; O'Neill, 2013, 2018). Instead, the theory states that, in pursuing one's interests, one must be able to justify one's actions to others with their own interests. We justify our actions by appealing to principles that no one could reasonably reject (Scanlon, 1998). This is T. M. Scanlon's version of contractualism, a more positive interpretation of which is that we justify our actions by appealing to principles that we could all agree to, or reasonably accept (Bagnoli, 2013). It is the basis of social agreement and cooperation in achieving collective goals. Given natural differences between the preferences of individuals and groups, the lower threshold of what no one could reasonably reject may be a more viable model for cooperation. Virtue theory may be compatible with contractualism in that the virtuous person's reasons for cooperation, and their appeal to principles, become a settled habit of the heart (Swanton, 2005, Chs. 3, 11; Adams, 2006). The virtues displayed in this type of behaviour are not of the theoretical sort associated with Kant or some contemporary philosophers but of the practical sort associated with Aristotle and our nature as social, or political, animals. Again, trust and fairness come to mind (O'Neill, 2018). Virtues are not just reflections of rational and moral agency but of a pluralistic set of individual and dispositions that constitute a good life.

The equal moral status of persons is the most basic of these principles. Fairness and trust are ideas based on equal moral status and mutual respect and in many ways are a further development of Aristotle's practical virtues. These are significant because the conditions necessary for people to flourish consist not only of reasons promoting tolerance but also reasons promoting acceptance and cooperation among all human agents for mutual benefit. One might reflect on the benefit to all of respect for Black American music. These reasons are objective in the sense that any rational agent can recognise and respond to them (unless 'you jus' don't get it'). But they do not correspond to any moral truths allegedly existing independently of us and human emotions. Rather, they are constructed by humans to meet the physical, cognitive, and affective demands placed on them by the natural world. In this regard, behavioural norms are not grounded in meta-ethics but in ethical naturalism. Foot's concept of natural goodness fits broadly within this framework.

Rejecting the metaethical view that moral progress depends on appeal to objective moral truths (Parfit, 2011), Philip Kitcher argues that individuals and societies engage in moral practices such as cooperation to address problems that beset them. They address, and in many cases solve, these problems through collective discussions that become more inclusive, better informed, and involve an increasing number of participants who are more inclined to engage with the perspectives of others and establishing behavioural norms that are acceptable to all (Kitcher, 2021). Kitcher's ethical naturalism aligns in many respects with Foot's conception of natural good. It is natural in the sense that behavioural norms are constructed and evolve according to how we live and act in the physical, social, and cultural environment. Moral behaviour depends on individuals and communities creating cultural environments in which we recognise and respond to the rights, needs, and interest of others. Moral progress depends on the extent to which this recognition and response occurs. The character of morality is not innate but a function of our capacity to create environments that are inclusive rather than exclusive and extend equal moral standing to all humans and some non-human animals (Buchanan & Powell, 2018; Buchanan, 2020).

We will further discuss Foot's conception later in this chapter. In the next two sections, we explain how the brain is a social organ and reflect on its role in mediating the neural and mental capacities necessary for moral reasoning and decision-making. While mental capacities are not reducible to neural-mental function, natural goodness presupposes a normal level of brain function.

THE SOCIAL BRAIN

Neurophilosophical naturalism is connected with ethical naturalism. Neural processes mediate the mental capacities necessary for the social interaction and norm-governed behaviour that allows moral progress by enabling more people to recognize and respond to the rights, needs, and interests of others. This claim does not, however, imply that morality can be explained entirely in terms of brain function. Instead, it points to the brain as a necessary but not sufficient factor in explaining whether or to what extent a person acts or fails to act in accord with social norms. This suggests that the brain is a social organ, or, as noted in Chap. 2, an organ that mediates interaction between individuals and the environment (Fuchs, 2018). Contextual features external to the brain are an essential part of its ecology. Brain functions associated with cognitive and emotional processing in the orbitofrontal cortex, anterior cingulate cortex, and anterior insular cortex enable social behaviours like trust and cooperation. Lower systems probably nudge things in directions inflected by empathy, unease, suspicion, and the other affectively toned attitudes.

Problem-solving of complex scientific or social problems is a skill of social cognition consisting in the ability to represent other people's minds and understand them by relying on these and other brain regions to simulate their behaviour within ourselves. "We understand ourselves by observing other people and their reactions to us" (Adolphs, 2003, p 176; Graziano & Kastner, 2011; Graziano, 2014). These reactions are enabled by neural networks that form representations of others affectively charged as part of a shared social space. Thus, how others react to us, and how we react to them, is grounded in both cognitive and emotional processes. These processes in turn are grounded in the prefrontal and limbic regions that underpin and subtly inflect our autopoietically shaped and complex higher functions.

Michael Graziano and Sabine Kastner explain that humans have specialised neural machinery that enables social intelligence. "The primary role for this machinery is to construct models of other people's minds thereby gaining some ability to predict the behaviour of other individuals" (Graziano & Kastner, 2011, p. 98). While this claim may appear to conflict with our claim in Chap. 2 that neurodynamics implies that our actions cannot be predicted, it is consistent with our claim if we interpret 'predict' as 'expect' and without any determinist implications. Graziano and Kastner add that "social perception is like spatial localization in respect of being dynamic and holistic. Social perception is not merely about constructing a model of the thoughts and emotions of another person but also about binding those mental attributes to a location" (p. 99). The brain's capacity to contextualise information from the world allows cognitive and emotional flexibility in how we respond to it. How we recognise and respond to others, and how the brain enables these mental capacities, depends on the physical and socio-cultural contexts in which we live and act (Bratman, 2014). There is also the reservation that the machinery is protoplasm with its own resonances or 'vibes'.

Yet there often are extenuating physical and social circumstances that can pose obstacles to a person's ability to simulate others' behaviour within themselves. Poverty, famine, hunger, and starvation from natural disasters and political upheavals forcing migration can cause severe psychosocial stress that can disrupt normal mind-brain interaction and the internal-external consonance necessary for this type of simulation. Indeed, at a more fundamental level, they can weaken or undermine the ability to adapt to a radically changing environment, much less pursue different goods. These changes may impair psychomotor, cognitive, affective, and volitional capacities necessary for adaptability. In the case of political or climate refugees, impairment in these capacities and associated internal-external dissonance may be more common among elderly refugees forced to live in an unfamiliar and unsafe world because they are less resilient and less flexible in their behaviour (Wexler, 2008, pp. 141ff).

In these circumstances, the question is not whether they can engage in reasoning about which actions with allow them to flourish, or which reasons for actions are acceptable to all agents. More fundamentally, the question is whether they can survive. A world literally or metaphorically on fire seems incompatible with an ideal world in which we cultivate intellectual and practical virtues in aspiring to and achieving individual and social goods. These practices presuppose a natural and physical environment that can provide the necessary conditions for them. These life-enabling and health-sustaining conditions are increasingly under threat, not only from natural disasters but also from the consequences of our actual imprudent, irrational, and immoral actions and the thoughts which occasion them.

The Moral Brain

In Chap. 2, we described different neuropathologies and how they can impair the psychomotor, cognitive, affective, and volitional components of agency. They can thereby adversely affect the neurodynamic soul and the way it enables adaptive behaviour and provides opportunities to flourish. Neural dysfunction can weaken the mental capacities necessary to recognise and respond to reasons for or against different actions and to engage in effective decision-making to attain various goods. Normal neural functions are normatively significant insofar as they mediate the mental states that make us appropriate candidates for attributions of responsibility, praise, and blame. These normal functions are also necessary for us to understand and act in ways that reflect a broad normative interpretation of 'good'. The components of agency are not compartmentalised but overlap and interact to a considerable extent.

Specifically, cognitive and affective processes interact in enabling rational and moral decision-making about both self-interested actions and actions that respect the interests of others. In psychopathy, emotional deficits such as lack of empathy can impair not only moral reasoning but prudential reasoning as well. The affected individuals make decisions that are not only harmful to others but also to themselves. Thus, both rational and moral decision-making are mediated mainly by connections between prefrontal and limbic regions in the brain (Bechara et al., 1999, 2000). Subcortical regions such as the striatum and cerebellum also have their role in planning and have a role in cognition which goes beyond their role in motor functions.

Moral reasoning and decision-making are therefore not localised in particular brain regions but are mediated by distributed cortical and sub-cortical networks. The neural circuits mediating these mental processes are domain-general rather than domain-specific (Decety & Cowell, 2015). It can thus be misleading to try to associate an action or pattern of behaviour with activity or circumscribed patterns of activity in a particular brain region. Even if there were a strong correlation between brain activity and a person's decision to act at a specific time, this would not imply that the activity caused the decision or action. It is possible that a conscious decision caused a change in unconscious brain activity, rather than the other way around. It cannot be known whether changes in blood flow or glucose metabolism displayed by fMRI or PET imaging cause, are caused by, or merely reflect brain activity and how they affect the mental process of forming and executing intentions in decisions and actions. The selective imaging is, as we have noted, based on subtractive contrast in blood flow where common activity between tasks is reckoned to obscure localisation.

Consider the anterior cingulate cortex (ACC). The ACC is critical for internal conflict resolution and the ability to choose between the felt appeal of courses of action. "An emerging theme in functional MRI studies is that lateral prefrontal cortex is a focal point for cognitive-emotional interactions, which have been observed across a wide range of cognitive tasks" (Pessoa, 2013, p. 132; Decety et al., 2012). Given these interactions, impairment in cognitive processing can impair affective processing, and vice-versa. This in turn can impair the ability to recognise and respond to reasons for acting that respect the rights, needs, and interests of other agents (Decety & Wheatley, 2015; Liao, 2016). By impairing this ability, neural and mental dysregulation can impair the cognitive and emotional capacity necessary to understand and appreciate the idea of goodness and activities that display it.

In some psychopaths, cortical-limbic and cortical-striatal circuit dysregulation associated with it may go some way toward explaining their lack of empathy and failure to recognise moral reasons for or against actions that harm others (Blair, 2007, 2013; Hosking et al., 2017). This dysregulation may prevent them from having a normal conception of good. The social and empathic dysregulation may also partly explain their maladaptive rational decision-making (Hosking et al., 2017). It may prevent them from having a conception of the good that is intrinsically other-regarding. In some people with impulse control disorders or addictions, dysregulation between the PFC and a hyperactive insula or nucleus accumbens in the brain's reward network may not impair their capacity to recognise moral reasons (Cima et al., 2010). But it may impair their capacity to respond to them in constructing a balanced response to a life situation. Still, neural abnormalities alone do not provide a complete explanation of their behaviour. Individuals with these disorders can recognise when others have been harmed by their actions and have the capacity to inhibit them, even if they fail to exercise it. This may give them some control of their actions and be quite compatible with occupation of a selective sociopolitical niche.

These examples illustrate how natural goodness, or more fundamentally moral sensibility, as part of an advanced form of adaptive behaviour, involves interactive mental capacities mediated by interacting neural processes. Moral behaviour is also mediated by optimal levels of neurotransmitters like dopamine and neuropeptides like oxytocin, though the latter has been associated with positive in-group behaviour and antagonistic out-group behaviour (Decety & Cowell, 2015). An excess or deficit of these substances and activity in the neural circuits with which they interact can adversely affect moral sensibility and more generally one's ability to understand and pursue the good. Aristotle's mean is the goal.

It is important to emphasise, yet again, that imaging displaying activity of neural circuits, neurotransmitters, neuropeptides, blood flow, and glucose metabolism can contribute to but does not provide a complete account of a person's rational or moral behaviour. Except in some cases of significant brain damage and dysfunction, imaging showing correlations between these neural features and our behaviour do not provide a causal explanation of how we think and why we act, or fail to act, in different circumstances. "If God had looked into our minds, he would not have been able to see whom we were speaking of" (Wittgenstein, 1953, #217). There can be no brain without mind and no mind without brain. Yet the content of our thoughts, including our attitudes about other agents and the social context within which we interact with them, are represented in the brain but are not located in it. Mental content includes representations of the social context and is underpinned by neurocognitive activity in the individual. The evolution of and motivation for morality are not just products of brain function but emerge from how the brain responds to the physical and mental challenges of living in the world (*Dasein*, if you will). Normal brain function mediating normal cognitive-emotional processing is necessary for us to understand the general concept of good and pursue different goods. But neither the concept nor our understanding of it is located in the brain.

Consistent with our neo-Aristotelian conception of the neurodynamic soul, 'good' is defined not just in terms of adaptability, positive moral judgments, or justification of our actions. It includes a broader range of functions mediated by the brain and mind that we perform well, activities through which we achieve a level of human excellence. This may include teaching, music, scientific investigation, friendship, and other activities and callings. They are different types of excellence, in this broad sense of *arete*. Regarding Aristotelian characteristic functions, what matters is not only what we do or are motivated to do, but also how we do it.

GOODS AND GOODNESS

Foot loosely follows Elizabeth Anscombe in acknowledging Aristotelian categoricals or norms that a biological species must obey in its behaviour if it is to flourish (Anscombe, 1958). Such norms transcend the divisions of 'prescriptivism' and 'descriptivism' and apply whether or not the individual feels the inclination to follow them. One might call them norms internally and naturally related to being a creature of a certain type. Human beings experience the joy and pain of intentional action in ways related variously to such norms, sometimes to their detriment. For example, one sees restricting one's food intake in response to obesity as good. But doing so in response to imagined obesity is pathological. We coin a disease name for that disturbed and pathological way of going on— "*anorexia nervosa*". Thus, the sense of 'ought' here is categorical but holistically, and not merely electively, tied to a robust conception of human nature. Its discursive inflections may result in one wilfully acting other than in a good way. Akratic, or weak-willed, actions can also become pathological behaviour if one performs them repeatedly. This is at odds with the Aristotelian model of following a physiological, statistical, or scientifically specified descriptive norm to advance individual and collective flourishing.

Having identified a natural form of goodness that is not merely descriptive, Foot goes on to examine claims for individual freedom or nonconformity to social norms. Against even these norms, a person can stand out as an individual. One might even say that to do so was, in itself, part of the form of humanity. In this regard, she reminds us of Wittgenstein's deathbed affirmation that he had had a 'wonderful life'. Given his many trials and deviancies, this may seem puzzling, as reported by his friends who witnessed his affirmation (Malcolm, 1965, p. 64). But Foot reminds us of the many ways that we limit our choices by, for instance, binding ourselves with promises to fortify ourselves against the weaknesses of temptation, self-interest, and 'all in' forms of weakness of will in acting irrationally against one's own best interests (Pears, 1984).

Foot speaks of human goods such as the joy of living well and a kind of deep happiness that is rooted in basic things such as "home, and family, and work, and friendship" (2001, p. 88). She thus distances herself, albeit subtly, from some hedonists. Foot notes that self-esteem of an admirable and grounded kind resides in these things (p. 91). There is a robust conceptual connection between this idea and human flourishing among and with others that distances it from self-satisfaction. It might (in an 'all things considered' dysfunctional form) characterise a more self-absorbed person whose personality is scarred, and thus both disfigured and fibrosed (or sclerosed) by unfortunate personality tendencies. A deep love for another might notably be transformative for such a person.

Pleasure and pain are not irrelevant in such an account, far less excluded from it. But they are given their proper and not a pre-eminent place among other aspects of what makes up a truly good and sincerely affirmed human life. This is the kind of thing that might be sincerely witnessed among others at some public event centred on a person. These things might be 'quirky', fondly remembered, shared with a sense of affectionate and unashamed acquaintance, a rejoicing in privileged and admirable familiarity, or affirmed with a sense of 'bonhomie' at some celebration or special occasion.

Foot reflects our metacognition as human individuals living an examined life:

> For there is a way in which a good person must not only see his or her good as bound up with goodness of desire and action, but also feel that it is, with sentiments such as pleasure, pride, and honour. (p. 90)

She considers 'immoralism' in an instructive way particularly for a neuro-ethics that takes seriously the world view of indigenous thought. These often emphasise the deep and inextricable connections between individuals and groups and between them and the environment. In this regard, she (as do they) consciously departs from much contemporary moral theory resting entirely on principles to which we appeal to justify our actions. Goods such as friendship with our fellow human beings and even the animals with which we live rest on these connections and responses. It is through them that we become fulfilled in our desires and intentions. This departure from principilism does not imply that good or sound principles are obstacles to human fulfilment. Rather, they are insufficient but necessary steps on the way to realising it. We do not "throw them away" like the rungs on the ladder, and the ladder itself, as in Wittgenstein's *Tractatus* (6.54). Principles inform the goodness that emerges from a *hexis* of acting in accord with them.

Merleau-Ponty sees the human mind as holistic and imbued with somatic resonance so that we ought to find there the clues to the nature of the good for human beings. Sexual being illustrates the varied ways in which we might dismember this aspect of our being into the visual, the somatosensory, the affective, the relational and so destroy the organic harmony that is part of personal existence in its authentic and inclusive nature (1945/1962, p. 185). The whole in its inclusive neurocognitive sense and undissected by the intellect is truly and joyfully part of being human. In its dissected, and thereby disrupted, form it is pathological in the variety of ways that Freud and others have discussed at length (Gillett, 2009). These distortions disrupt our lives so as to render them disharmonious and riven with stress in many different ways. We classify only some of them as 'diseases', and even note that fewer of them than we might think have a pathophysiological cause (in the reductive sense). Sexuality, as Merleau-Ponty saw it, fits into the way of being human that enables other goods as a source not only of satisfaction and pleasure but also as a waypoint on the journey to belonging and community.

These goods are truly satisfying and enduring, not only immanent in but also outlasting and transcending human lives and mortality. Those internal or philosophical connections are often neglected in favour of scientifically demonstrable causes beloved of the self-images of this post-enlightenment age. This work aims to redress that balance in preparing for a more organic and natural conception of the good for human beings.

This enlightenment and post-industrial misdirection springs from some individually nuanced combination of conceptions of human excellence in our age. One might construct a series of contrasts that require resolution by a philosophical synthesis. One example is rhythms of life, which are open and dynamic. These can be contrasted with executive functions in the mind and brain, which, according to deterministic models, are fixed and predictable. But these functions do not follow a defined set of demands, a method or strategy of execution, and an outcome. The claim that they do fails to appreciate the realities of ethology. Rhythms of life form and develop organically in neurocognitive attunement to a context that takes on a definitive shape. A species has a mode of being with its own neural foundation in its organismic mode of being. Thus: "If a lion could talk, we could not understand him" (Wittgenstein, 1953, II, 190). That is, words have meaning, and can only be understood, in particular contexts of discursive interpersonal relationships. These are all forms of life. Human individuals are interestingly and importantly individuals, but not on that account less human in their being. The intentions we express in words can only be ascertained, or at least inferred, from real human relationships as real people in social interactions.

We must therefore consider the ways in which we are 'human-all-too-human'. This will turn out to involve football, dancing, music, and poetry as much as food and water and sexuality. The goal of these activities is to establish and maintain harmonious relationships and skills rather than to produce material outcomes. These relationships and skills themselves are valuable.

They are both interpersonal and intergenerational and embody techniques and associated cognitive abstractions or concepts which are attuned to balance, stability and fruitful innovation. We share with the lion family and community and collective endeavours in which both males and females take part. These complex interwoven activities can surprise our social norms as the she-wolves of Kipling and the shield maidens and women guerrillas of human warfare have shown. We should not be surprised, as Athene of mythology prefigured them. Even today we spin myths that disrupt our stereotypes such as 'The Devil wears Prada'.

Good lives are more than disconnected pleasurable or good experiences. They are Integrated ways of going on in exercising skills and experiencing their effects extended over all stages of life. These actions and their consequences promote human well-being by forming sustainable modes of adaptation and harmony within the ecosphere. This is a measure of the level of physical and mental functions that humans value. *Eudaimonia* is a state that results from having and exercising these functions at a high level. Although this is an overlap with the translation of eudaimonia as 'happiness', it also tends to undermine a definition of the latter as pleasure when it is construed merely hedonistically.

A good life comprises a number of ways of conducting oneself at both theoretical and practical levels. This multiplex task gives most of us (to some degree from trivial to immense) satisfaction. In so doing, our techniques of personal relationship, social conduct, and self-configuration and thus the totality of ways of being human is expanded as each of us is singular with a name that functions as a singular (or denoting) term. Cognitively we can meld together dialogue and reflection as part of this mix. The former exposes us to new ways of thinking or revives for us forgotten ways of mental activity. The latter encourages a synthesis that introduces a person to a more inclusive and challenging way of forming the self and his or her ideas and dispositions.

This results in our each forming a unique way of occupying a place in the world about which we can feel good and sustain a certain self-respect. It will serve us well throughout life and guide our dynamic interactions with others. The alternative is to harden or fossilise in a pre-formed world view. Some who consolidate former views and dwell in them without the stasis of fossils serve a valuable function, reminding us of cognitive paths that can be trodden without the distorting influence on experience of preformed fixity in or conformity to the self-images of the age. The first poignantly reminds one of an insect preserved in amber that once was alive enough to get stuck in plant sap as a result of its own appetites when the life-extinguishing event occurred.

Just as an insect can be drawn by its appetites into a cloying death-trap of tree-sap, a person might also be so drawn because of primal and infantile forces in the psyche and be drawn into paralytic states of the psyche such as hysteria. These are insightfully discussed by Merleau-Ponty and linked to the contrast between freedom and resentment (1945/1962, p. 189). Freedom allows us to be open to experience and grow. Resentment cripples and distorts the psyche, twisting its participation in life and particularly in life with others. The latter distorts our self-knowledge so that Descartes' 'cogito' becomes insecure in a way that is unsettling for much of the discourse of philosophy. It is interesting in this regard that P.F. Strawson should entitle a collection of his oft-neglected essays, *Freedom and Resentment* (1974), and in it consider and problematise many hallowed doctrines in philosophy of mind. These include causation in perception, social morality, freedom of the will, the structure of language, and self, mind, and body. In each case, such 'hallowed' doctrines are critiqued, and apparent distinctions called into question (so they do not hollowly and solemnly ring in the literature). So, one turns "*tae* think again" (although Strawson hardly grounds the Scottish allusion). In its totality, this is what becoming a good, or well-functioning, human being comprises. Given its genetic and epigenetic provenance, it is not only a natural but also an intentionally attained complex of dynamic functions that may not be clearly seen until the end of the process.

Being good must not traduce these many complex and intersecting thoughts about human nature in favour of a tendentious philosophical doctrine, whether it be a variation on descriptivism, expressivism, emotivism, or any other 'ism'. Nor should it be identified with Parfit's 'triple theory' consisting of Kantianism, consequentialism, and contractualism as the culmination of normative and meta-ethical inquiry (Parfit, 2011). It is not based on rules prescribing certain types of conduct. Goodness, on the present account, becomes an intrinsic part of the content of consciousness, which is holistic and embodied and involves subjective openness to the world and others. This openness is in constant balance with the considered interests of self towards its own appetites and desires. These are perfectly good in themselves as part of the greater and more all-encompassing personal balance within which all the things that we enjoy play a proper and healthy part, actually and conceptually.

This is a more modest conception of goodness than the one described by Mark Johnston:

> The good person is one who has undergone a kind of death of the self; as a result he or she lives a transformed life driven by entering imaginatively into the lives of others, anticipating their needs and true interests and responding to these as far as is reasonable. The good person is thus a caretaker of humanity…. (Johnston, 2010, p. 14)

While natural goodness in knowing how to go on includes anticipating others' needs and interests, it does not involve a death of the self but a cultivation of it. One does not lose oneself in others. Nor is one their caretaker. Natural goodness does not entail a type of altruism where one is always or generally motivated to act for the benefit of others (Batson, 2011; Pfaff, 2015). Rather, one cultivates the self both by pursuing one's interests and by engaging with others and their interests in undertaking and achieving individual and collective goals.

Aristotle noted the type of natural goodness we have described and the way it moderated human desire. A fullness of life is achieved against the things that would distort it by overly focusing on some one thing. Thus, in ethics we might aim for and prescribe the Aristotelian mean. This balanced focus of aspiration could include aspects of being human which range from individualism to sociality, from pacifism and tolerance to bellicosity and violence, from tolerance to prejudice and egotistical self-worship. In each of these behaviours, there is a point of balance and proper regard as against a tendency to gravitate towards an extreme, and the balance requires right thinking and a seasoning of humility. That point or integrative aim with a modicum of reflection brings to mind the song "Poppa don't preach" (by Madonna—the one who is *not* the mother of God).

We might wonder what it is that has engendered the astigmatism in favour of technological and manufactured achievement by contrast with reflection on nature and an acceptance of the restraint required to allow us to continue to co-exist with things. This is a relation relatively unspoiled by our intervention or restored where appropriate. We might mention *hubris* and a delight and fascination in our own cleverness, so that blade art is dismissed as being impractical, even though prominent in the archaeological record. One is reminded of the works of Saruman and Mordor of recent mythology (with ancient roots). Of course, the world *can* be organised, systematised, and 'made to order'. But that order is not ab-original and may not rightly sense ancient balances, as the extinction of the *moa*, among so many other species, might highlight.

We are re-making our world in both life-giving and life-damaging ways. Discerning the difference is not easy. The birth defects in children born from women prescribed thalidomide to relieve nausea during pregnancy comes to mind. At some point we all, like 'proud Edward', must turn back to think again. Judging that point will take inclusive wisdom in which the dominant and conquering ideology (made to order in every sense) must learn to take its proper place.

We are natural, evolved, creatures who have genetically and epigenetically developed cognition to the point of it being an elaborate combination of language, meaning, and thought. This combination enables us to reflect upon and revise our ways of going on so that they equip us for a natural and artefactual world, dynamically changing as we interact with it. In this way, we modify the context in which we interact and develop our own capabilities of perceiving and acting. We thus transform the dynamic reality that is human existence into a holistic reality that is partly of our own making. We must constantly learn anew what is good in that context, seeing the past clearly so that we can properly learn to see our way ahead. Neurophilosophy is but a particular way of exploring some of the intellectual demands associated with that task.

Conclusion

Being good consists in functioning well across a range of intellectual and practical activities. It is not something that we intuit, grasp, or understand by appeal to alleged objective facts or truths that exist independently of humans. Instead, it is behaviour that emerges when we act and interact with others in ways that allow us to desire, pursue, and achieve individual and collective goals. There is a multiplicity of goods defined in terms of physical functions that organisms perform at an optimal level for bodily homeostasis, and there are also cognitive and emotional functions that humans perform at this level for flexibility, adaptability, and the capacity for achievement and a new world. The general concept of natural goodness includes all these functions. Human functions associated with natural goodness correspond roughly to the Aristotelian concepts of the human characteristic forms of life that include rational activity (*oi nou*), excellence in executing cognitive, emotional, and motor skills (*arete*), practical wisdom in knowing how to act (*phronesis*), and flourishing or fulfilment (*eudaimonia*) that emerges from them. There is a natural evolution from the first to the last of these activities. Goodness includes but is not limited to actions that affect others in positive ways, or the motivation for these actions. Goodness consists not only in *what* we do but also *how* we do it.

Thus characterised, the form of life involved is not just a moral concept involving behaviour that conforms to rules, or that consists in discharging duties and obligations. Nor is it just a moral concept based on social agreement about principles of action that all can accept. Goodness consists in the 'knowing how' that is expressed by going on in a range of activities and actually living well in light of this knowledge rather than trending with a new 'Age of stupid'. Particular goods include friendship, cooperation, or any activity performed at a highly skilled level in harmony with nature and its enduring balances. Achieving these goods requires cognitive-emotional skills and a collective wisdom that can resolve and move beyond conflicting individual and cultural points of view (Kane, 2013).

Desiring and achieving natural goods presupposes a certain degree of consonance between one's psyche and the setting in which one lives and acts. One's moral character depends on the right social environment for realizing one's moral potential (Buchanan, 2020). But this environment can change Extenuating circumstances in hostile physical and social environments can present challenges in adapting to the world. This durable consonance, by contrast, also presupposes a normally functioning brain that mediates the psychomotor, cognitive, affective, and volitional capacities necessary to adapt to unexpected changes in the environment and thrive or flourish in it in a way that aligns with our desires, beliefs, and intentions. Neural oscillations and synchronization are influenced by the brain's response to the environment and each other. The dynamic brain mediates this engagement and our interaction with other agents. But the brain alone does not provide a complete explanation of this engagement— the natural goodness characteristic of a range of human functions and skills performed well. Goodness is an emergent property of a holistic and integrated set of relations between and among the brain, mind, body, and world.

References

Adams, R. M. (2006). *A theory of virtue: Excellence in being for the good*. Clarendon Press.

Adolphs, R. (2003). Cognitive neuroscience of human social behavior. *Nature Reviews Neuroscience, 4*, 165–178.

Anscombe, G. E. M. (1958). Modern moral philosophy. *Philosophy, 33*, 1–19.

Arola, A. (2011). Native American philosophy. In J. Garfield & W. Edelglass (Eds.), *The Oxford handbook of world philosophy* (pp. 562–573). Oxford University Press.

Bagnoli, C. (Ed.). (2013). *Constructivism in ethics*. Cambridge University Press.

Batson, C. D. (2011). *Altruism in humans*. Oxford University Press.

Bechara, A., Damasio, A., & Damasio, H. (1999). Different contributions of the human amygdala and ventromedial prefrontal cortex in decision-making. *Journal of Neuroscience, 19*, 5473–5481.

Bechara, A., Damasio, A., & Damasio, H. (2000). Emotion, decision-making and the orbitofrontal cortex. *Cerebral Cortex, 10*, 295–307.

Blair, J. (2007). The amygdala and ventromedial prefrontal cortex in morality and psychopathy. *Trends in Cognitive Sciences, 11*, 387–392.

Blair, J. (2013). Psychopathy: Cognitive and neural dysfunction. *Dialogues in Clinical Neuroscience, 15*, 181–190.

Bratman, M. (2014). *Shared agency: A planning theory of acting together*. Oxford University Press.

Buchanan, A. (2020). *Our moral fate: Evolution and the escape from tribalism*. MIT Press.

Buchanan, A., & Powell, R. (2018). *The evolution of moral progress: A biocultural theory*. Oxford University Press.

Cima, M., Tonnaer, F., & Hauser, M. (2010). Psychopaths know right from wrong but don't care. *Social Cognitive and Affective Neuroscience, 5*, 59–67.

Decety, J., & Cowell, J. (2015). The equivocal relationship between morality and empathy. In J. Decety & T. Wheatley (Eds.), *The moral brain: A multidisciplinary perspective* (pp. 279–302). MIT Press.

Decety, J., Michalska, K., & Kinzler, K. (2012). The contribution of emotion and cognition to moral sensitivity: A neurodevelopmental study. *Cerebral Cortex, 22*, 209–220.

Decety, J., & Wheatley, T. (Eds.). (2015). *The moral brain: A multidisciplinary perspective*. MIT Press.

Foot, P. (1972). Morality as a system of hypothetical imperatives. *Philosophical Review, 81*, 305–316.

Foot, P. (2001). *Natural goodness*. Oxford University Press.

Fuchs, T. (2018). *Ecology of the brain: The phenomenology and biology of the embodied mind*. Oxford University Press.

Gauthier, D. (1986). *Morals by agreement*. Oxford University Press.

Gillett, G. (2009). *The mind and its discontents* (2nd ed.). Oxford University Press.

Ginsborg, H. (2020). Wittgenstein on going on. *Canadian Journal of Philosophy, 50*, 1–17.

Graziano, M. (2014). *Consciousness and the social brain*. Oxford University Press.

Graziano, M., & Kastner, S. (2011). Human consciousness and its relationship to social neuroscience: A novel hypothesis. *Cognitive Neuroscience, 2*, 98–113.

Hobbes, T. (1651/2017). *Leviathan* (Ed. C. Brook). Penguin.

Hosking, J., Kastman, E., Dorfman, H., et al. (2017). Disrupted prefrontal regulation of striatal subjective value signals in psychopathy. *Neuron, 95*, 221–231.

Johnston, M. (2010). *Surviving death*. Princeton University Press.

Kane, R. (2013). *Ethics and the quest for wisdom*. Cambridge University Press.

Kant, I. (1785/1964). *Groundwork of the metaphysics of morals* (Trans. H. J. Paton). Harper & Row.

Kitcher, P. (2021). *Moral progress*. Oxford University Press.

Korsgaard, C. (2008). Aristotle's function argument. In *The constitution of agency: Essays on practical reason and moral psychology* (pp. 129–150). Oxford University Press.

Liao, S. M. (Ed.). (2016). *Moral brains: The neuroscience of morality.* Oxford University Press.

Lindemann, H. (2019). *An invitation to feminist ethics.* Oxford University Press.

Maitra, K., & McWeeny, J. (Eds.). (2022). *Feminist philosophy of mind.* Oxford University Press.

Malcolm, N. (1965). *Ludwig Wittgenstein: A memoir.* Oxford University Press.

Merleau-Ponty, M. (1945/1962). *Phenomenology of perception* (C. Smith, Trans.). Routledge.

O'Neill, O. (2013). *Acting on principle: An essay on Kantian ethics.* Cambridge University Press.

O'Neill, O. (2018). *From principles to practice: Normativity and judgement in ethics and politics.* Cambridge University Press.

Parfit, D. (2011). *On what matters,* volumes I and II. Oxford University Press.

Pears, D. (1984). *Motivated irrationality.* Clarendon Press.

Pessoa, L. (2013). *The cognitive-emotional brain: From integrations to integration.* MIT Press.

Pfaff, D. (2015). *The altruistic brain.* Oxford University Press.

Rousseau, J.-J. (1761/1968). *The social contract* (Trans. M. Cranston). Penguin.

Scanlon, T. M. (1998). *What we owe to each other.* Harvard University Press.

Stewart, G. T. (2021). *Māori philosophy: Indigenous theory for Aotearoa.* Bloomsbury.

Strawson, P. F. (1974). *Freedom and resentment and other essays.* Methuen.

Swanton, C. (2005). *Virtue ethics: A pluralistic view.* Oxford University Press.

Taborsky, M., Cant, M., & Komdeur, J. (2021). *The evolution of social behaviour.* Cambridge University Press.

Wexler, B. (2008). *Brain and culture: Neurobiology, ideology, and social change.* MIT Press.

Wittgenstein, L. (1921/1974). *Tractatus Logico-Philosophicus* (D. Pears & B. McGuinness, Trans.). Routledge.

Wittgenstein, L. (1953). *Philosophical investigations* (G. E. M. Anscombe, Trans.). Macmillan.

EPILOGUE

Current neurophilosophy, with its origins in European and Anglo-American thought, labours under enlightenment abstractions and lacks a post-colonial voice. These abstractions are replete with metaphors that misrepresent human neurocognition and generate metaphysical and logico-mathematical paradoxes. This has prompted some to return to sources such as Aristotle as a founding voice of non-dualistic psychophysical naturalism. However, attempting to distil holistic neurocognitive engagement with the ethosphere into one or more syllogisms will only generate other puzzles or contradictions. We have used syllogisms in this book. But they should be understood as analogous to what Wittgenstein says about the propositions in the *Tractatus* (6.54). We need them to express philosophical positions. But we also need to leave them behind when we understand the multifactorial nature of being in the world. Different claims arise in different distillations of meaning, and it is no wonder that they conflict from time to time. Only an obsession to possess a pure and incontestable 'Truth' could tempt us to try to resolve every paradox rather than humbly live with our many cognitive shortcomings.

Phronesis, *arete*, and *eudaimonia* indicate an Aristotelian route to a conception of human natural function in the world we inhabit. The first of these is knowing how to do things by being in command of a range of techniques in goal-oriented behaviour. The second is a good, well-functioning, or excellent character, living within our limitations and

G. Gillett, W. Glannon, *The Neurodynamic Soul*, New Directions in Philosophy and Cognitive Science, https://doi.org/10.1007/978-3-031-44951-2

accepting them while striving to do better in an inclusive way with others. The third is living well, or flourishing, through these techniques and efforts. It is the natural outcome of the first two. These three states are grounded in the Aristotelian conception of soul (*psuche*) as a psychophysical process. Consistent with this tradition, our conception of the neurodynamic soul is a unified psychophysical process consisting of neural rhythms that influence and respond to the organism's body and mental states and enable it to engage with and adapt to (or adaptively resonate with, as in ART) the natural, social, and cultural environment.

The 1986 Argentine film *Man Facing Southeast*, directed by Eliseo Subiela, depicts a patient in a psychiatric hospital. He asks his psychiatrist if he can assist with autopsies of deceased patients in the pathology lab. While dissecting tissue from a cadaver's brain, he asks the pathologist: "Doctor, where was the soul of this man?" In the spirit of Wittgenstein and our conception of the neurodynamic soul, one could reply that the person's soul was not located in his brain but was an emergent property of neural rhythms and how they and the mental states they generate and sustain are shaped by the context in which a person lives, acts, and interacts with others.

Rats and their reactions to operant conditioning have much to teach us. Rats are omnivores like us, only much more cognitively limited. Their discontents, however, can be illuminated by this type of conditioning and teach us much about ourselves. They can be increasingly driven by various contingency regimes and their associated stimulus conditions, as can we. But those who were Skinner's disciples sometimes showed the dogmatism of many true believers. Indeed, their inconsistency with each other in such complexes of behaviour, motivation, and reaction to stimuli can teach us much about ourselves, as anyone involved in child psychology can attest. In particular, the harm caused by aversive contingency regimes and 'catch 22' contexts suggest that this is a broadly encompassing problem in the animal kingdom which must be overcome if we are to thrive. The children of those among us trapped in poverty, abuse, deprivation, and deculturation attest to the monetary or other social evils we inflict on one another in the service of this or that ideology or regime.

Being *unheimlich* reminds us that an increasingly common human experience is of displacement, marginalisation, homelessness, and not being 'at home' in one's ways of going on. At home in a caring family, one is properly surrounded by love and concern for one's well-being. That concern is differently manifest as one develops, and a personality forms on a combined genetic and epigenetic basis. One develops the techniques of living at home as a site of self-formation. Such self-making ideally occurs in a culture amidst a family who share love and, where it can be gently communicated in love, truth (which may be hard to hear about oneself and even harder to heed). Once formed, we are each set free to live and apply our diverse cognitive and affective resources to a broad range of activities. In the real world, we are set free into a discursive context where we must learn to live (*disce vivere*).

Dynamic adaptation to a biopsychosocial mode of being in a context creates a person's way of being, or their mastery of techniques, which are well or ill adapted to our being in the world with others. Logico-mathematical structures may be an extension of neurocognition. But they at most provide a partial picture of the human mind-brain. Picture theories of this relation always come up short. These structures may tempt us to see life as an abstract set of propositions. It is much more than this.

Flexible culturally inflected being embodies the self-images of the age. It is consistent with neuroscience and yields a more inclusive view of human thinking and acting. Given the normative aspects of these activities, this type of being also offers a more inclusive view of ethics. Thus, neuroscience is both located within science and liberated from its causal chains. It is located in demonstrable science and dynamic functional neuroimaging. Yet it is liberated from strict causality by the power of human imagination and stories that engage us and, partly through our affirmation, configure us.

Being discursive allows us to imagine and execute life plans on the basis of a broad neurocognitive context. This context includes counterfactuals and forms of simulation about different future possibilities and courses of action. It allows us to take on ideas derived from individual and collective imagination and to resist cramming the air with promises. This 'diet' allows us to be chameleons indeed, unrestricted by a 'Diet of worms'.

The ethical significance of consciousness and intention is that we engage in moral action with wide and growing choices. These choices are not precluded by some combination of natural laws and events in the past but are potentially open to us by the contingencies of lived experience. The choices we make result in actions or words which make a difference to the world. They both reflect and affect culture and human lives There is thus a responsibility inextricably bound up with such freedom; but this is a complex matter on which we have tried to shed some light (apologies if the phraseology borders on the kind of pedantry 'up with which', Winston Churchill said, 'he would not put').

Consider the abstract quasi-mathematical likelihood of a particular British bullet hitting a French soldier at the battle of Vitoria in 1813, given some of the following possible causal contributory factors. It is a particular British bullet (of many more fired at the time with a similar intention. Muskets are very inaccurate and fired in volleys, so aimed at a body of troops; rifles may contribute to the volley). Now consider possible qualifiers:

(a) It is a rifle bullet—many less fired, each more likely to hit a target because more likely to be fired by a rifleman and therefore hit its intended target.
(b) The French soldier is an officer—many less in the battle, much more likely if it is a rifle bullet as riflemen target officers, more likely if the officer is carried away by patriotic fervour and 'leads from the front'.
(c) That still does not address questions of range, hastiness, distractions, discipline, individual skill, coolness under fire, etc, etc.

This is not quite the tortuous reasoning and double guessing of "The Princess Bride" poison scenario. But it is a form of human reasoning struggling to distil a pure linear causal story out of a complex real-world and historical event, a story limned in general terms or as comprising particular 'actions and events'(apologies to Davidson). Every neural network moment (temporal not mechanical) is like that. Two major features of the account emerge.

1. Including a postcolonial voice in current neurophilosophy reinforces the suggestion that post-industrial metaphors, when applied to brain function, misrepresent human neurocognition. Neurodynamic conceptions of human cognitive function have prompted a return to Aristotle as a founding voice for non-dualistic naturalism. His best interpreters (e.g. Charles, 2021) in contemporary conceptions of human thought seem to have a more fluid conception than what is evident in mechanistic and reductive paradigms.
2. Philosophy tends to forget the fact, much discussed by Plato, that many abstractions, if interpreted literally, lead to an impasse. For instance, in the dialogue with Euthyphro he develops an argument

that the same act can be both 'pious' and 'impious' if there are multiple Gods and they disagree (an assertion that follows from 'The Iliad'). We might consider this an argument more broadly applicable to religion (especially in light of European wars and the global nature of learning in academia). This argument should not set us against reflections inspired by the human spirit. But it should warn us against an over-emphasis on solidifying doctrine into dogma in a process that divides us as peoples rather than unites us in a search for reaffirmation of humanity with a *soupcon* of mystery that endures despite attempts to explain it away.

This vision goes beyond 'dreams and visions' to the lives of those down-to-earth and physical folk less enslaved by cognitive astigmatism.

Consider a practical syllogism based on geometry and a sporting skill:

(a) A ball hit straight hits its mark, hit otherwise (curved or crooked) it is less likely to do so and a greater element of chance enters in.
(b) A skilful football player can spin a ball so that it travels in a curved path to a designated end point.
(c) A straight ball can be blocked or deflected from its target.
(d) A ball can be curved so as to hit its target and elude interception.

Such is a skill or technique about which one can think so as to become a goal-scoring forward. But the abstraction a (with the added lemma a1—'a curve is a kind of crookedness') gives the synthesis c by adding b2 'a curve might coincide in end point and starting point with a straight line'. The practical outcome of this reasoning is a goal-scoring skill able to be mastered as a technique. Cognition alone cannot yield this mastery. It requires a physical and not merely mental technique. Like all techniques, mastery takes practice. This has philosophical implications beyond football or baseball. The apparent mathematical 'a' is modified by 'the touch of the real, and the luck (*tuche*), or contingencies, that accompany it. This results in a *phronimos* characterised by prudent and skilled behaviour.

The perspective we have outlined can be developed in as many ways as human creativity can devise, such as music or dance. Creativity involves a modified human adaptive context alive with imagination, and we indwell it whatever its historical or geographical location. In the first part of the twentieth century, Black musicians became 'at home in their sounds' (R. Gillett, 2021), ultimately giving rise to a two-way symbiosis between

Paris and the USA. This helped to undo much White American conde-scension about things 'Black'. That momentous interwar (first and sec-ond) cultural change partly drawing on 'comradeship in arms' was not confined to those states who had recognised the relevant moral principles, and it was far from universally recognised, accepted, and inwardly digested even there. Nevertheless, it became an integral part of American *phronesis*, *arete*, and *eudaimonia* in the *psuche* of modern culture (albeit attenuated by racism in some jurisdictions).

Inclusive adaptation to the biopsychosocial context restores people to their way of being (Bolton & Gillett, 2019). But it must be nurtured. Wittgenstein describes this human adaptation as comprised of mastering techniques, which is particularly important for people who are forming themselves. Still, being in the world as a distinct and highly developed neurocognitive creature can have adverse consequences if one forgets one's limitations. These limitations are part and parcel of being well-adapted. Logico-mathematical explanations of mental structure as a con-ception of neurocognition can result in over-immersion in a metaphor and can give us a false picture of the human mind-brain. Other developments of that view, and rarefied forms of it hold similar perils for conceptions of human thinking.

As discussed in Chap. 2, Libet claims that the human brain acts prior to forming a conscious intention. From this, he concludes that the conscious intention was not part of the genesis of the action. However, we might argue that the soul, or the holistic neurocognitive core of a human being, is in continuous integrated adaptive interaction with the environment. Particular actions are snapshots of that dynamic reality with the aim to isolate and individuate the action from its proper woven conceptual place within the broadly integrated cognitive unfolding of the individual's life. To isolate such an event in this seamless activity is to fall under the distort-ing spell of the doctrine of 'brain events' beloved of certain neuro-philosophers in thrall to the post-industrial model of mind and brain.

Aristotle, the classical model for natural philosophers, laboured under no such handicap (Gillett, 2018). For that reason, perhaps later philoso-phy has focussed critically on the abstract cognitive arguments of Plato or the clear-cut metaphysical distinctions of Descartes. The later syntheses of Husserl and his followers and the critical notes struck by Nietzsche and Wittgenstein have been sidelined by many as lacking rigour. Early neuro-philosophical voices such as Hughlings Jackson, Bergson, or Merleau-Ponty have not been heeded by many contemporary philosophers. Indeed, some consider them as surpassed by those who, departing from Freeman and neurodynamic research, divide neurophilosophy into empirically valid post-industrial neuroscience and unscientific speculation. The rhythms of the brain give the lie to that excursion into post-industrial metaphor.

Vygotsky and then Luria escaped due recognition by many because they transcended the prevailing metaphors of the day. They were outside the most prominent centres of scientific and politico-cultural power for the early part of their careers. Luria came later to be recognised for his hugely important and integrative approach spanning laboratory and field studies, confounding many in the age of neural localisation. His work was centred in London and Boston but had axes (shared with Hughlings Jackson) radiating widely out from Queen Square and Harvard.

The many philosophical critiques of Libet's main claim about brain activity preceding conscious intention would have been more compelling if they had recalled this seminal work. The real history of thought precluded them from appealing to neurodynamics and phenomenological and post-modern theory. Informed by clinical, anthropological, and experimental evidence, these theories allow a more inclusive *ubersich*, whereby any lived moment is seen as part of a situated human story. Neural activity gives rise to intentional actions and multifaceted experiences variously integrated into a life lived partly through and generated by one's neurocognitive stream of experience. These experiences in turn influence neural activity at specific times and over time.

Libet's experimental design was well-suited to explore the precise time relations between discrete causally related physical events. But it was not well-suited to the more dynamically related moments of the actual neurocognitive stream. Any abstraction from the continuously adapting interaction between the organism and the environment artificially divides that stream into mental events and their physical counterparts. But spatially and temporally discrete causally related mental and physical events are not fluid biological moments entwined with the dynamic reality of real life in the world in which we live and move and have our being.

An animal digests a legume it has eaten. But at what moment and through what causal event does the animal digest it? It is misguided to try to break up the digestive stream into simple events when it is woven together with processes and holistic associations of the neurocognitive flow of life and ways of going on. It is true that neurocognitive 'events' seemingly occur at temporal 'moments'. On a linear causal account, an act at T3 is preceded by an intention at T2 and a neural event at T1. Yet although actions take shape intentionally in time and culminate in an intended result if well executed in a timely way, there is many a 'slip between the cup and the lip'. Intentions may miscarry on the way to the actions they are designed to produce. Multiple factors inside and outside the brain can influence the outcome. The brain activity preceding successful or miscarried agency does not occur as discrete events and cannot be abstracted from the mental character that shapes and is shaped by it.

Flexible culturally inflected being embodies the self-images of the age and is realistic in light of neuroscience. It turns our thinking towards a more inclusive view of human thought and action. It suggests a more holistic view of ethics beyond the idea of prescriptive thought and towards the good conceived as the fitting understanding and aim of life for any creature. Flexible being includes self-images in a reflective gaze that eschews conceptualising any one of them as exclusively 'True'. Truth is multi-faceted and can be viewed from many cognitive and conative perspectives. What may look 'thus' from one perspective may reveal itself to be 'so' from another, especially when all possible and actual life situations are considered.

Being discursive allows human beings to imagine and execute life plans on the basis of all these broader neurocognitive moments and processes. These can be accessed from differently interested and positioned *loci* of being affecting both time and momentum. Thus, they can disclose a conception of the good which is inclusive and not limited to a set of current contingencies. It is not quite 'a view from nowhere'. It is, rather, a view with a *whakapapa* and has a discursive origin watered by *aroha*, the 'gentle rain from heaven'.

Discourse fuelled by *aroha* is active and pulsates with intersecting rhythms of life among people. We might here invoke three levels of complexity:

1. The complex environment changing moment-by-moment in innumerable ways.
2. The neural network interacting both with the living body and with a culture in its environmental context.
3. The complexity of the genome modified dynamically through epigenetics.

Consciousness, intention, and action emerge from this interaction as it is articulated by discourse. They are all part of a shared natural language, communicated, and full of adjectival and adverbial nuances. These nuances are different aspects of what philosophers such as J.L. Austin described as how we do things with words (Austin, 1962; Searle, 1969; Fogal et al., 2018). Austin explores the complex interaction between language and the world in ways that are deeply informed by life and human creations and institutions and tend to confound logico-mathematical formulations. Following Wittgenstein, this is philosophy which "leaves everything as it is" (1953, I, 124), rather than pursue a more draconian project of analysing or trying to explain every aspect of our experience.

Here we find a problem for the philosophy of consciousness similar to what arises when we inquire into the meaning of language. Just as one cannot answer a question about 'red' by indicating a red patch unless it is clear that the question concerns colour, not shape, brightness, or salience of visual presentation, so one needs more cognitive differentiation or direction to provide a sensible answer to questions about the meaning of other terms and concepts. Language comprises many elements. Words often have multiple meanings, depending on context. 'Singular' bears a mathematical and expressive meaning, drawing on a thing being remarkable or the centre of attention in some way.

The woven universe of human thought intrigues and sometimes captivates us in its 'weft' and 'weave' of integrated human lives. As part of this fabric, meaningful acts using language vary hugely with the situations in which they occur. For example, two women looking at hats and one says, "My dear it's the Taj Mahal". Or Winston Churchill, pointedly responding to a woman chastising him for being drunk "Yes Madam, and you are ugly; but in the morning I will be sober". In each case, there is a wealth of meaning in the words that a 'flat-footed' or literal reading would miss. A logico-mathematical abstraction, divorced from the contexts in which they were used, would also miss their meaning These actual discursive moments show that language and understanding are inextricably entangled with life. They indicate that a pure philosophy of language modelled on logic and mathematics is an artificial beast of human invention at odds with the natural lifeworld of human neurocognition.

When we range into dynamic neurocognitive science, we find some provocative concepts. 'Attractors' is a term used to describe neural rhythms in human organisms. They are one way of neurocognitive going on that develops in the life story of an organism in an ethological context. The context in which the individual acts shapes the meaning of the act is usually rich indeed. ART is significant in this regard in describing the brain's balance between processing new information and maintaining stable representations of the world. This theory subtends a similar *double entendre* to QBism/Cubism. Both terms bring to mind the resonant interaction between neural rhythms, human thought, and how we act in and interact with the natural and social world. We need to learn to pulse with neural processes symbiotically rather than consumptively and to metaphorically avoid the threat to us as agents that the eponymous disease actually did in scarifying our lungs and bodies. The present pathology of trying to explain being in the world in abstract or reductive terms involves a scarring of both body and soul. (Ironically, Katherine Mansfield dies of TB).

Illusions, hallucinations, dreams, and delusions arise sometimes through disordered minds. Sometimes they arise through 'language on holiday'. Real cognition and real work is demanding but also rewarding, as the unemployed and also the depressed, or mentally dissociated will attest. We cannot afford that type of 'shared insanity' or 'holiday' infecting and affecting our normal way of being because it neither satisfies nor sustains us. Our good is to construct a world and live together in an enduring way so that our descendants will say, with more enduring justification than when it actually was said, "This was their finest hour".

This is an evaluative thought and so concerns the good. The ethical significance of consciousness and intention is that we engage in moral action with wide and growing choices which can make a positive difference to the world. That is done in proper cognisance of both what is real and what is counterfactual. Choices are thrown up or forced upon us by an embodied and embedded life as human beings at some place in some stretch of lived historical time. In that sense, there is 'no exit' except the eternal self-denial of suicide (or 'sui-genocide'). This is both an indictment of and challenge to us all. It carries varying degrees of responsibility, depending on the human circumstances involved and how we respond, or fail to respond, to them.

This book has explored that thinking and those circumstances and the way that human neurocognition has dealt with them through our history and currently deals with them. That history, *inter alia*, lives in striking ideas that have emerged and are still emerging among those of us who contemplate the self-images of the ages. Our analysis and discussion of the neurodynamic soul reveals some of the limitations of established philosophical conceptions of mind and brain in which many have found comfort. These limitations challenge all of us to move beyond them in a constructive way that more accurately captures how we think, act, and live.

We live in the shadows of giants, and the trick is neither to be happily at rest in the darkness nor blinded by the light.

REFERENCES

Austin, J. L. (1962). *How to do things with words*. Clarendon Press.

Bolton, D., & Gillett, G. (2019). *The biopsychosocial model of health and disease: New philosophical and scientific developments*. Palgrave Macmillan.

Charles, D. (2021). *The undivided self: Aristotle on the mind-body problem*. Oxford University Press.

Fogal, D., Harris, D., & Moss, M. (Eds.). (2018). *New work on speech acts*. Oxford University Press.

Gillett, G. (2018). *From Aristotle to cognitive neuroscience*. Palgrave Macmillan.

Gillett, R. (2021). *At home in our sounds: Music, race, and cultural politics in interwar Paris*. Oxford University Press.

Searle, J. (1969). *Speech acts: An essay in the philosophy of language*. Cambridge University Press.

Wittgenstein, L. (1953). *Philosophical investigations* (G.E.M. Anscombe. Trans.). Macmillan.

INDEX

© The Author(s), under exclusive license to Springer Nature Switzerland AG 2023
G. Gillett, W. Glannon, *The Neurodynamic Soul*, New Directions in Philosophy and Cognitive Science,
https://doi.org/10.1007/978-3-031-44951-2